THE ORIGINS OF
KANT'S ARGUMENTS
IN THE
ANTINOMIES

THE ORIGINS OF
KANT'S ARGUMENTS
IN THE
ANTINOMIES

———

SADIK J. AL-AZM

OXFORD
AT THE CLARENDON PRESS
1972

Oxford University Press, Ely House, London W. 1

GLASGOW   NEW YORK   TORONTO   MELBOURNE   WELLINGTON
CAPE TOWN   IBADAN   NAIROBI   DAR ES SALAAM   LUSAKA   ADDIS ABABA
DELHI   BOMBAY   CALCUTTA   MADRAS   KARACHI   LAHORE   DACCA
KUALA LUMPUR   SINGAPORE   HONG KONG   TOKYO

PRINTED IN GREAT BRITAIN
AT THE UNIVERSITY PRESS, OXFORD
BY VIVIAN RIDLER
PRINTER TO THE UNIVERSITY

PREFACE

In this study I have tried to shed light on Kant's arguments in the antinomies of pure reason by relating them to certain aspects of the historical intellectual background out of which the critical philosophy grew.

The philosophy of Kant has been carefully treated by great masters indeed. Consequently, I can hardly hope to do more than contribute some improvement to our understanding of the origins of Kant's reflections and writings about the subject of the antinomies.

I am indebted to my friend and colleague Mr. David Makinson not only for his constant encouragement and stimulation but also for the many valuable criticisms, corrections, and insights with which he supplied me after reading the manuscript in its entirety.

S. J. A.

Beirut, Lebanon
April 1971

CONTENTS

THE FIRST ANTINOMY

THE importance of the problem of antinomies in decisively affecting Kant's thought in the very early stages of the development of the critical philosophy has been generally recognized by scholars for some time.[1] In a long letter to Marcus Herz dated 21 February 1772, Kant explained that his preoccupation with the problems of the antinomies of pure reason had a decisive influence in setting his mind on the path of the critical philosophy.[2] Again, in 1798, Kant wrote to Christian Garve to inform him that his starting-point in the elaboration of the critical teachings was the problem raised by the antinomies.[3] Kant himself described the impact of the problem of the antinomies upon philosophy in the following words: 'It serves as a very powerful agent to rouse philosophy from its dogmatic slumber and to stimulate it to the arduous task of undertaking a critical examination of reason itself.'[4]

[1] Concerning the details of this question see de Vleeschauwer's article 'Les antinomies kantiennes et la *clavis universalis* d'Arthur Collier', *Mind*, New Series, 47, 1938, 303–20. See also G. Martin, *Kant's Metaphysics and Theory of Science*, Manchester University Press, 1955, p. 42.

[2] *Kant: Philosophical Correspondence 1759–99*, translated and edited by Arnulf Zweig, University of Chicago Press, 1967, pp. 70–9. According to Norman Kemp Smith, 'Benno Erdmann has very conclusively shown, preoccupation with the problem of antinomy was the chief cause of the revolution which took place in Kant's views in 1769, and which found expression in his *Dissertation* of 1770'. (*A Commentary to Kant's Critique of Pure Reason*, Humanities Press, New York, 1950, pp. 431–2.)

[3] *Kant: Philosophical Correspondence*, p. 252.

[4] *Prolegomena*, pp. 337–8.

It would be safe, I think, to recognize an autobiographical reference in this statement in the sense that the problems raised by the antinomies influenced Kant's mind in the same direction as did his acquaintance with certain aspects of the philosophy of David Hume. All this indicates the strong affinities of the antinomies with the early statements of the critical point of view as we find them elaborated in such works as the *Inaugural Dissertation* of 1770 and the *Transcendental Aesthetic*.

The First Antinomy deals with problems relating to space and time. To avoid repetition I shall limit myself in this discussion to the question of space. The thesis of the antinomy informs us that the world is limited as regards space (i.e. it is spatially finite), while the antithesis states that the world has no limits in space (i.e. it is spatially infinite). Kant proves the thesis by assuming the antithesis in the body of the proof and showing it to be false. Then he assumes the thesis in the body of the proof of the antithesis and proceeds to show that it is also false. In other words, he demonstrates both thesis and antithesis by resorting to indirect proof.[5]

Before proceeding to examine the arguments presented by Kant in favour of the thesis and antithesis I should like to note the following interesting comment made by T. D. Weldon on the antinomies:

---

[5] It will be significant for us to note at this point that this ancient method of argumentation is constantly employed by Clarke and Leibniz in their famous controversy. The two disputants kept refuting each other by assuming the major claims of the opponent and then proceeding to demonstrate their falsity and absurdity in a variety of ways. See, for example, Leibniz's fourth letter, paras. 16 and 17, and his fifth letter, para. 65. (*The Leibniz–Clarke Correspondence*, ed. H. G. Alexander, Manchester University Press, 1956.)

It is immediately clear that Kant considers the theses to be the *a priori* contentions of rationalist cosmology, while the antitheses represent the empiricist attack on it, and also that the truth of the theses rather than that of the antitheses is desirable both on practical and speculative grounds.[6]

This, however, is far from being immediately clear. If Weldon's contention is to stand, then it must be carefully argued. Indeed if we wish to introduce rationalism and empiricism into the discussion, then the exact opposite of what Weldon claims appears to be true. In other words, it is the thesis which is closer to empiricism, while the antithesis is very much in the spirit of rationalist metaphysics and cosmology (Leibniz and Wolff).

I shall try to point out that the thesis really expresses the position of the Newtonians in the form in which it was defended by Clarke in his famous controversy with Leibniz. And the Newtonians were certainly closer to the experimentalist spirit long dominant in English science and philosophy than the rationalist cosmologists of the Continent. For example, Clarke often used to appeal, in his dispute with Leibniz about the reality of *vacua* in nature, to conclusive empirical evidence in favour of his position on this question to which Leibniz usually retorted by means of arguments based on purely speculative and *a priori* considerations.

The argument for the antithesis is ultimately based on the grounds utilized by Leibniz and his followers to reject the Newtonian conception of space and time with its cosmological, philosophical, and theological implications.

In fairness to Weldon and to those who hold similar

[6] T. D. Weldon, *Kant's Critique of Pure Reason*, Oxford University Press, 1958, p. 204.

opinions concerning this question[7] I should mention that Kant, in exploring the *extra-rational* reasons and motives which lead thinkers to rally around the thesis or the antithesis as the case may be, does identify the antitheses of his antinomies with 'empiricism' and their theses with 'dogmatism' (A465–76, B493–504). However, careful scrutiny of the relevant texts will show that Kant is using the terms 'empiricism' and 'dogmatism', in this context, in a quite unusual way which makes their meaning very different from the purport of the terms 'empiricists' and 'rational cosmologists' as they occur in Weldon's statement. Kant's salient example of an 'empiricist' here is Epicurus and not some orthodox representative of empiricism in the usual sense. His example of the 'dogmatist' is Plato (A471–2, B499–500). This choice of examples should be sufficient, in itself, to warn the reader that Kant is not attributing to 'empiricism' and 'dogmatism' their usual significations when he identifies them with the antitheses and theses of the antinomies respectively.

When Kant identifies the antithesis (i.e. the assertion that the world is spatially and temporally infinite) with 'empiricism' he means to say that those philosophers who adhere to this point of view reject the enterprise of

[7] For example, E. Caird identifies the antithesis with an empiricism 'which often slides into dogmatic materialism'. But the truth of the matter is that it is the Galilean–Newtonian conception of a finite amount of matter distributed in infinite absolute space that has supplied the basis and main support for modern materialism. In other words, it is the thesis (which incorporates in itself the 'fallacy of simple location') that has been historically identified with materialism and not the antithesis. (*The Critical Philosophy of Immanuel Kant*, Glasgow, 1889, ii, 50.) Views similar to Caird's are also held by: F. Paulsen, *Immanuel Kant*, Scribner's, New York, 1902, pp. 215–16, and James Ward, *A Study of Kant*, Cambridge University Press, 1922, p. 36. See also H. W. Cassirer, *Kant's First Critique*, Allen & Unwin, 1954, pp. 273–82.

searching for the absolute beginning, end, and cause of our universe. They hold instead that because the universe is infinite no such unconditional states of affairs can ever be reached. The 'Empiricist' (in this particular sense) '. . . will never allow, therefore, that any epoch of nature is to be taken as the absolutely first, or that any limit of his insight into the extent of nature is to be regarded as the widest possible' (A469–B497). On the other hand, those who hold to the thesis (i.e. the belief in the finitude of the world) are called 'dogmatists' because they seek an intelligible beginning for the universe, and maintain 'that all order in the things constituting the world is due to a primordial being, from which everything derives its unity and purposive connection' (A466–B494). It should be clear then that 'empiricism' in this sense is characteristic of the philosophical views of many rationalist cosmologists, and that 'dogmatism' in this sense is characteristic of the philosophical beliefs of many orthodox English empiricists. Strictly speaking, the conflict represented in the first antinomy is not between empiricism and rationalism, as Weldon claimed, but between two kinds of strict dogmatism. In fact Kant himself characterizes the antithesis as dogmatic also (A521–B549 n.). Consequently, this conflict, Kant tells us, cannot be successfully 'mediated' save by an appeal to the critical temper of the mind.

However, this does not mean that upon reading the formal statement of the antithesis(es) and its proof one does not emerge with the overall impression that a 'naturalistic' position is being stated here. That is a position somewhat different from the official Leibnizian metaphysics with its spiritual atoms, confused perceptions, etc. Therefore we should note here that, although the ideas expressed in the four antitheses are identical in essentials with the views adopted by Leibniz and are constantly supported by Leibnizian arguments, it is the Leibniz of the correspondence that we are primarily dealing with and not quite that of the *Theodecy* (for example). What I mean is that some of the views Leibniz expounded and some of the assumptions he accepted in the correspondence about the nature of space, time, material substance, etc., do not fully harmonize with his formal metaphysical doctrines. The reason for this is simple and to some extent tactical. In conducting the controversy with Clarke, Leibniz was naturally restricted by its limited context and scope. He also imposed restrictions upon himself in the interest of furthering the cause of the debate. The context of the topics under discussion was of a scientifico-philosophical nature whereby the subject is treated primarily at the level of 'phenomena' and not so much at the level of the ultimate metaphysical realities constituting the universe. The limited context of the controversy with Clarke is hardly the appropriate place for Leibniz to elaborate his metaphysical doctrines. That would have certainly opened him to the charge of irrelevance and of dragging in considerations that do not speak directly to the specific point under sharp debate. Furthermore, Leibniz's task was to refute the arguments of his opponent the way they were presented (primarily at the level of phenomena), not simply to produce his own metaphysical point of view and present it alongside with, and as an alternative to, that of his opponent. This would hardly make for a refutation or even a debate. Leibniz, accordingly, granted tentatively, and at times solely for the sake of the argument, a number of assumptions and points which depart from his official metaphysical views as stated elsewhere. Thus, if we limit ourselves to the correspondence, the Leibniz which emerges grants (at least implicitly) the reality of extended substances, the ubiquitous application of natural law and the reality of spatial relations among extended substances. This Leibniz hardly mentions the monads as the true atoms of things or the confused nature of sense-perception. In other words, it is a sort of a 'naturalistic' Leibniz which we encounter in the correspondence; this is clearly reflected in the views and arguments of the antitheses and their proofs, as will become evident in due course.

Another point I would like to clarify here is that my argument does not mean to imply that Kant was simply a good historian of the great intellectual and scientific controversy between Leibniz and the Newtonians. However, the spectacle of this conflict must have provided him with one of the finest and most striking instances of what an 'antinomial conflict' could be. Kant takes into account the contrasting claims and arguments of Leibniz and Clarke in a manner adapted to his basic philosophical intention of combating metaphysical dogmatism and of serving the broad interests of the critical philosophy. Consequently, in constructing his four antinomies, Kant abstracts from the details and concrete circumstances surrounding the controversy and generalizes the intellectual confrontation to make out of it a basic problem of pure reason as such. Thus according to Kant the antinomy is 'not arbitrarily invented but founded in the human reason as such' and 'all the metaphysical art of the most subtle distinction can not prevent this opposition . . .'.[8]

---

[8] *Prolegomena*, pp. 337–8, 339–40.

We should note in connection with the thesis and its proof, that what is being asserted as finite is the material universe regarded as 'a given whole of coexisting things' *in* space and not space itself.[9] From the point of view of the thesis the question of the finitude or infinitude of the world involves only the material universe found in space (and time) and does not encompass the spatial container itself. The spatial container is assumed to be infinite and the thesis specifically claims that the totality of coexisting things in space are not coextensive with all the points of space. Kant explains in the proof of the antithesis that 'if the world in space is finite and limited' then it 'exists in an empty space which is unlimited'. Another important implication of this view is the asser-

[9] This important distinction has been neglected by most writers on the problem of the first antinomy. They consider the issue at stake between the thesis and antithesis to be the infinitude or finitude of space itself and not of the world in space only. For example, George Schrader writes about the proofs in the first antinomy: 'The argument attempts to show that space and time must be both finite and infinite and since this involves a self-contradiction, they cannot be real entities'. ('The Transcendental Ideality and Empirical Reality of Kant's Space and Time', *The Review of Metaphysics*, iv, 4, 1951, p. 532.) Again, Norman Kemp Smith often leaves the strong impression, in his discussion of the thesis of the first antinomy, of neglecting to draw this same distinction. He writes as if the thesis is meant to assert the finitude of space and time as well as of the world in space and time. (*Commentary to Kant's Critique of Pure Reason*, pp. 483–8.) Again, H. A. Prichard implies in his treatment of the subject that the contradiction underlying the first antinomy pertains to 'the actual infinity of space and time'. (*Kant's Theory of Knowledge*, The Clarendon Press, Oxford, 1909, p. 101.) A. C. Ewing avoided this error and recognized that both the thesis and antithesis presuppose the infinitude of space and time. (*A Short Commentary on Kant's Critique of Pure Reason*, Methuen & Co., London, 1950, p. 211.) G. Martin went beyond Ewing and pointed out the importance of Newtonian and Leibnizian cosmological ideas in the construction of the first antinomy. (*Kant's Metaphysics and Theory of Science*, pp. 42–6.)

tion of the actual existence of empty space beyond the region where the world is supposedly located in space.

Now, I hardly need to emphasize that the ideas expressed in the thesis are not characteristic of the rationalist cosmologists of the Continent as Weldon claims, but are straightforward statements of the Newtonian position as it was expounded and defended by Clarke in his letters to Leibniz. In fact the observation on the first antinomy leaves little doubt that the thesis is meant to state the Newtonian point of view. Kant writes: 'But while all this may be granted, it yet cannot be denied that these two non-entities,[10] empty space outside the world and empty time prior to it, have to be assumed if we are to assume a limit to the world in space and in time' (A433–B461).

The argument of the thesis may be summarized in the following manner:

1. Assume that: 'the world is an infinite given whole of coexisting things'.

2. The magnitude of a quantity which is not given to intuition as limited 'can be thought only through the synthesis of its parts'.

3. It follows that in order to think the totality of such a magnitude (or quantum), we need to complete, 'through repeated addition of unit to unit', the process of synthesis.

4. It also follows that in order to think the infinite world as a whole the process of 'the synthesis of its parts' must be completed. But by the very nature of an infinite magnitude no process of the synthesis of its parts can ever be completed.

[10] Kant at times refers to absolute space and time as two non-entities by way of criticizing the Newtonians. See A39–B56.

5. Therefore, the world is finite because:

An infinite aggregate of actual things cannot therefore be viewed as a given whole, nor consequently as simultaneously given (A428–B456).

In order to elucidate the details of this argument we should seek some explanation of what Kant means by a 'world' and the relationship it has to the process of 'synthesis'. In the opening paragraph of his *Dissertation* of 1770 Kant writes:

Just as, in dealing with a complex of substances, analysis ends only with a part which is not a whole, i.e., with the *simple*; so synthesis ends only with a whole which is not a part, i.e., with a world. (35)[11]

We noted that in the proof of the thesis Kant speaks of 'an aggregate of actual things' and applies to it such descriptions as: 'simultaneously given', 'a given whole of coexisting things', and 'viewed as a given whole'. Kant seems to think, here, that if the aggregate of things forming the material universe is to form 'a whole which is not a part', i.e. a *world* in the strict sense (or a *totality* in the language of the *Critique*) then the synthesis of its parts ('through repeated addition of unit to unit') must end at a point where the aggregate is 'viewed' as a 'given whole of coexisting things'. In order to give a less abstract and more imaginative account of Kant's thoughts on this question I shall resurrect Laplace's famous image of a superhuman intellect which looks at the universe through the medium of Newton's *Principia* and sees its entire past, present, and future in one glance.

[11] All such numbers refer to the pages of Kant's *Inaugural Dissertation and Early Writings on Space*, translated by John Handyside, Open Court Publishing Co., Chicago, 1929.

Accordingly, Kant's expression of 'a given whole of co-existing things' (and its equivalents) would mean that the aggregate of things forming the material universe can be said to constitute a *world* if and only if we can conceive of it as an instantaneously present fact to the gaze of such a superhuman intellect. That Kant might have had such an image in his mind is indicated in the *Dissertation* where he says:

... an intellect is possible which might apprehend an aggregate distinctly at a glance, without the successive application of a measure, though such an intellect would not indeed be human. (37 n.)

It is, therefore, necessary for the human intellect, in conceiving this aggregate as a 'totality' (i.e. as a *world* or whole), to resort to the process of synthesis because no appeal can be made 'to limits that of themselves constitute it a totality in intuition'.[12] Kant is, in effect, reasserting his distinction between an analytic totality and a synthetic totality. In the former case we proceed from an intuition of the whole to an analysis of the parts which are conditioned by the given whole itself. Space and time are considered analytic totalities in the *Aesthetic* and *Dissertation*. Therefore, Kant speaks of space 'as an infinite *given* magnitude' whose parts are *within* and not *under* it (B40). On the other hand, the material universe is conceived in the thesis as a synthetic totality because in this case we cannot 'proceed from the whole to the determinate multiplicity of the parts', but we 'must demonstrate the possibility of a whole by means of the successive synthesis of the parts'.[13] It is interesting to note in this connection that the

[12] [13] Observation on the thesis.

concept of infinitude utilized by Kant in this argument bears a striking resemblance to Locke's famous explanation of the idea of infinity. Kant explains the concept of infinitude that he is applying in the following words:

The true transcendental concept of infinitude is this, that the successive synthesis of units required for the enumeration of a quantum can never be completed.[14]

From this he concludes:

. . . the concept of totality is in this case itself the representation of a completed synthesis of the parts. And since this completion is impossible, so likewise is the concept of it.[15]

Locke's account of the infinitude of magnitudes runs as follows:

Every one that has any idea of any stated length of space, as a foot, finds that he can repeat that idea: and joining it to the former, make the idea of two feet . . . and so on, without ever coming to an end of his additions . . . after he has continued his doubling in his thoughts, and enlarged his idea as much as he pleases, he has no more reason to stop, nor is one jot nearer the end of such addition than he was at first setting out. . . . This, I think, is the way whereby the mind gets the idea of infinite space.[16]

From this Locke concludes about the idea of infinity:

. . . it is nothing but a supposed endless progression of the mind, over what repeated ideas of space it pleases; but to have actually in the mind the idea of a space infinite, is to

[14] Observation on the thesis.

[15] Observation on the thesis. Kant makes the same point in the *Dissertation*: 'In consequence neither analysis nor synthesis will be completed in such a manner that the former will yield the concept of the simple and the latter the concept of the whole, unless each process can be brought to a conclusion in a finite and assignable time' (36).

[16] *Essay*, bk. ii, ch. 17, para. 3.

suppose the mind already passed over, and actually to have a view of all those repeated ideas of space which an endless repetition can never totally represent to it; which carries in it a plain contradiction.[17]

Locke and Kant are in agreement, then, that the concept of a synthetic totality which is also infinite is a contradiction in terms. For Kant only an analytic totality can be infinite and if the material universe is a synthetic totality (a *world*) then it is necessarily finite.

In the light of these explanations, the crux of Kant's argument for the thesis may be recast in the following fashion:

(*a*) On the supposition that the aggregate of things in space (the material universe) is infinite, then it can never form a *world* (totality or whole) because the synthesis of its parts can never be completed.

(*b*) But the actual aggregate of things does form a *world* (or a totality), the synthesis of its parts can, in principle, be completed and viewed by Laplace's intellect as a 'given whole of coexisting things'. It forms a synthetic totality.

(*c*) Therefore the material universe (the *world*) is finite. Obviously (*b*) is the silent premise underlying the entire argument.

It should be clear, therefore, that the argument of the thesis works to draw some of the implications of the particular conception of the *world* (totality, whole) presupposed in the argument. Given this definition of a *world* it follows necessarily that if an aggregate of substances forms a *world* then it is finite. Given the silent premise of the argument we infer that since the actual

[17] *Essay*, bk. ii, ch. 17, para. 7.

universe is an aggregate of substances which form a *world* then it is finite.

One might be tempted, at first, to object that Kant produces no reasons why we should accept this particular definition of a *world* and why we should assent to the assertion contained in the silent premise (that the actual material universe does in fact form a world). P. F. Strawson raised such an objection against Kant in connection with certain points pertaining to the problem of time in the first antinomy.[18] But in fact no such reasons are needed since Kant's definition of a *world* is really a reassertion of a central claim of the classical Newtonian conception of the universe in terms that are distinctive to the critical philosophy. The reasons for accepting Kant's definition of a *world* are the same reasons that compelled generations to accept, almost unquestioningly, this Newtonian claim. The claim I am referring to is Newton's inclusion of position in absolute space in the definition of a physical (material) entity. Accordingly, such an entity has to be finite or else it will fill all spaces, i.e. it will have no determinate position relative to absolute space which amounts to a violation of its own definition. From the point of view of the thesis the physical entity called the universe requires, then, by definition, a determinate position in absolute space, i.e. it cannot fill all spaces. It has to be finite. Furthermore, the immense success of the classical science of motion led Newtonians to the natural inference that the actual universe is in fact constructed precisely after the fashion of the definitions, laws, and principles just enunciated by the great master of the science of motion. In other

[18] *The Bounds of Sense: An Essay on Kant's Critique of Pure Reason*, Methuen & Co., London, 1966, p. 178.

words, the silent premise underlying the argument of the thesis is an expression of the inescapable, and often unarticulated, belief of Newtonians that the actual material universe is Newtonian in nature and character. In the terminology of the thesis it forms a *world* or a totality. Consequently, Kant cannot be expected to produce reasons in favour of accepting this definition of a world and the silent premise following it since he was not trying to argue here in favour of any such principles. On the contrary, he labours to construct the antinomy in order to expose the illusoriness of all such 'metaphysical' claims that are in principle irresolvable. Kant's task consists in 'presenting for trial the two opposing parties, leaving them, terrorised by no threats, to defend themselves as best they can, before a jury of like standing with themselves, that is, before a jury of fallible men' (A476–B504). His rendering of the thesis and its argument is based on the well-known metaphysical views (on the subject of the spatial magnitude of the material universe) of the Newtonians.

Failure to discuss the first antinomy in relationship to the background of Newtonian ideas and conceptions that it expresses has led Strawson to level additional unfair criticisms and objections against Kant's presentation of the argument of the proof of the thesis. Commenting on this argument Strawson says in his essay on *The Critique of Pure Reason*:

To this argument it can be objected, first, that the hypothesis of an infinite spatial extent of the *world* does not require the possibility of the completion of a temporally infinite process of spatial surveying of the world. For that possibility requires at least the further hypothesis that the world has existed for an infinite time.[19]

[19] *The Bounds of Sense*, p. 177, emphasis is mine.

One need not quarrel with Strawson's statement as such. However, its value as an objection to Kant's argument in the thesis depends primarily upon the meaning of the term 'world'. Obviously Strawson does not attribute to it the strict Newtonian sense which occurs in the thesis (and proof) of the first antinomy. Since the thesis is really a forceful and faithful statement of the Newtonian point of view and its implications concerning this matter, the objection becomes irrelevant. The kind of objection that one can relevantly bring against Kant in this context is to question the accuracy of his statement of the implications of the Newtonian concept of a *world*. The reply to Strawson's objection consists then in simply pointing out that the conception of the *world* according to the thesis entails the absurd result (from the point of view of the thesis) that the 'hypothesis of an infinite spatial extent of the world' will necessarily require 'the possibility of the completion of a temporally infinite process of spatial surveying of the world' on account of the fact that the *world* forms a totality. And Kant explains:

The concept of totality is in this case simply the representation of the completed synthesis of its parts; for, since we cannot obtain the concept from the intuition of the whole [as in an analytic totality]—that being in this case impossible—we can apprehend it only through the synthesis of the parts viewed as carried, at least in idea, to the completion of the infinite.[20]

Another example of Strawson's often historically irrelevant criticisms of Kant on these questions is:

. . . it appears that in declaring the completion of an infinite temporal process to be impossible, Kant is repeating

[20] Second footnote to the proof of the thesis.

the mistake which we have found in his argument for the first part of the thesis. A temporal process both completed and infinite in duration appears to be impossible only on the assumption that it has a beginning.[21]

Kant is not committing a mistake here at all. He is simply expounding the implications of the Newtonian conception of a *world* which is based on the firm assumption that every process has a beginning in infinite receptacle-time. Consequently, no such process can be completed and infinite at the same time. One can hardly blame Kant for raising these philosophical questions and formulating their implications in terms of the ideas and conceptions that resulted from the greatest scientific achievement of his century.

Again, failure to dicuss the first antinomy against its background of Newtonian ideas and conceptions has led Norman Kemp Smith to criticize Kant's conceptions of a 'whole' and of finitude as they occur in the thesis. Kemp Smith writes in criticizing Kant:

The use of the words 'given' and 'whole' is misleading. If space is infinite, it is without bounds, and cannot therefore exist as a whole in any usual meaning of that term. For the same reason it must be incapable of being given as a whole. Its infinitude is a presupposition which analysis of actually given portions of it constrains us to postulate, and has to be conceived in terms of the definition employed in thesis *a*.[22]

There are two points to be noted in connection with this criticism:

(*a*) I have tried to point out in the previous pages that Kant does not employ the term 'whole' in any usual

[21] *The Bounds of Sense*, p. 177.
[22] *Commentary*, p. 485.

meaning of that term but in a strictly defined and specified sense. Kant took the trouble in the *Dissertation* and supplementary comments and observations on the thesis to indicate the meaning to be attributed to the term 'whole' (or totality) in this context. Consequently there is no justification for Kemp Smith's assertion that the use of the term 'whole' is misleading.

(*b*) From Kant's definition of a 'whole' as always finite Kemp Smith infers that on these grounds space cannot form an infinite whole, contrary to earlier Kantian claims to this effect. But Kant's definition of a 'whole' in the thesis applies strictly to the world in space and not to space itself. Failure to draw this distinction leads Kemp Smith to apply what can be properly said of the *world* in space to infinite space itself, and the result is naturally a confusion for which Smith is responsible and not Kant. The 'whole' we are concerned with in the thesis is a synthetic totality while space forms an 'infinite whole' only as an analytic totality. Kemp Smith can conclude that space 'must be incapable of being given as a whole' only if he construes 'the whole of space' as a synthetic totality as he ought not to do. After all, Smith himself emphasized the early composition of the section on the antinomies and its strong affinities to the early statements of the critical teachings in the *Dissertation* and *Aesthetic*. But in spite of this fact he regards the infinitude of space in the thesis as a conceptual inference from actually given portions of space, i.e. as a synthetic totality. This conception of the infinitude of space is not characteristic of the earlier stages of the critical philosophy and arises only with the *Analytic*. For the *Dissertation*, *Aesthetic* and the first antinomy:

(*a*) space is 'an infinite given magnitude'[23] and not a concept.[24] It is an analytic totality. (*b*) The *world* is the 'concept of a whole which is not a part' arrived at through the synthesis of its parts. It is a synthetic totality. Kemp Smith's criticism is based on a confusion of these two distinct interpretations of the nature of a whole.

Kemp Smith criticizes Kant's statement in the proof of the thesis that the world cannot be infinite because on such an assumption it will be impossible to complete in thought the successive synthesis of its parts. According to him Kant is here inferring '*an objective impossibility of existence*' from '*a subjective impossibility of apprehension*'.[25] On these grounds he accuses Kant of committting the fallacy of *ignoratio elenchi* for failing to distinguish 'the objective conditions of existence' from 'those of mental apprehension'. This criticism would appear plausible only if we dilute the Locke–Kant account of infinitude to hold that because we can never, in a finite time, subjectively form an idea of an infinite totality by the progressive synthesis of finite parts, then such a totality cannot exist in actual fact. But in spite of Locke's and Kant's psychological metaphors and subjectivist language their final conclusion on this point is not that an infinite totality cannot exist in fact because subjectively we are unable to form the idea of such a totality, but because such an idea is impossible in the sense that it is a self-contradiction. There is little to be

[23] B40.
[24] Kant states in the *Dissertation*: '*The concept of space is a singular representation*, including all spaces *in* itself, not an abstract common notion containing them *under* itself' (59).
[25] *Commentary*, p. 485.

gained by following Smith in giving the problem under consideration a subjectivist psychological twist. The problem of the antinomy is primarily a conceptual problem (or at least should be regarded so). I think that Kant makes it quite clear, as I have tried to show, that from the point of view of the thesis an infinite world is impossible not on account of the limitations of the human 'conditions of mental apprehension' but because the concept of a *world* logically entails the idea that the synthesis of its parts must take a finite time to complete irrespective of the contingent fact as to whether the human powers of apprehension can actually achieve such a feat.

Finally, I should like to allude to certain misleading characterizations that Kemp Smith makes in his treatment of the first antinomy. For example, he writes about the proofs of the thesis and antithesis: 'They are formulated in terms of the dogmatic rationalism of the Leibnizian position, with a constant appeal to abstract principles.' And '. . . Kant believed that the rationalistic proofs which he propounds in their support are unanswerable, so long as the dogmatic standpoint of ordinary consciousness and of Leibnizian ontology is preserved'.[26] These statements give the impression that somehow the thesis of the antinomy embodies the 'dogmatic rationalism of the Leibnizian position' or 'the dogmatic standpoint of ordinary consciousness'. It would be far more accurate to say that the thesis embodies the ideas and conceptions of the dogmatic rationalism that grew around the Newtonian conception of the universe. To this Kant opposed, in the antithesis, the idea and conceptions of the dogmatic rationalism of

[26] *Commentary*, pp. 481, 483.

the Leibnizian metaphysics as they appeared in the correspondence with Clarke. In this connection George Schrader has contended that the antinomy is invalid because 'contradictory conclusions about space and time are not demonstrated as required for a genuine antinomy'.[27] Schrader's contention would be correct if the antinomy were intended to demonstrate contradictory conclusions about space (time) itself. But, as I have already mentioned, Schrader is mistaken in holding that the antinomy is about the finitude and infinitude of space and time themselves. Once this misconception is avoided it becomes evident that Kant has succeeded in framing a genuine antinomy in the sense that he has put before us for consideration two dogmatic, mutually exclusive, internally coherent metaphysical claims about the *real nature of the world* with no prima facie means for a rational choice between them. What more can we require from Kant in order to grant that he has framed a valid antinomy! I shall return to this subject at a later point in this discussion.

Before I move on to consider the antithesis I would like to point out that on the basis of the previous discussion the first antinomy can be nicely correlated with the *Dissertation* of 1770 as a whole. It will be remembered that the *Dissertation* treats of 'the form and principles of the sensible and intelligible world'. Now according to 'the principles of the form of the sensible world', i.e. Newtonian space and time, the idea of an aggregate of substances forming an infinite *world* or whole is impossible. This is exactly the contention of the

[27] 'The Transcendental Ideality and Empirical Reality of Kant's Space and Time', *The Review of Metaphysics*, iv, 4, 1951, p. 532.

21

thesis of the first antinomy as I have tried to show in the previous pages. The part of the *Dissertation* dealing with the intelligible world is Kant's concession, at this stage of the critical teachings, to Leibnizian cosmology and rationalist metaphysics. According to 'the principles of the form of the intelligible world' the idea of an infinite whole is not only legitimate but even necessary.[28] And this is exactly the claim of the antithesis: 'the world is an infinite given whole of coexisting things'. The first antinomy is in this sense already contained in the *Dissertation*.

The argument of the antithesis may be summarized in the following manner: on the assumption that the world in infinite space is finite, then things will not only stand in spatial relationships to each other, but will also stand, as an aggregate, in a certain relationship to space itself. But since such a relation to empty space is a relation to no object it follows that it is nothing real. Therefore, the world is spatially infinite, i.e. it fills all space.

Taken at its face value, this argument is quite shabby. However, careful scrutiny of the footnotes and the observation on the antithesis will show that there is much more to the argument than at first meets the eye. Therefore, I shall try to point out in the subsequent pages that the proof of the antithesis, along with its supplementary explanations, represents fairly accurately the grounds on which Leibniz and his followers rejected the Newtonian conception of space (and time) with its cosmological, philosophical, and theological implications.

[28] See Robert P. Wolff, *Kant's Theory of Mental Activity*, Harvard University Press, 1963, pp. 12–13.

Speaking in general we may say that from the point of view of the antithesis no valid distinction can be drawn between an absolute spatial container, on the one hand, and the world existing in it, on the other, since according to Leibniz space (and time), without things, is only a mere ideal possibility.[29] Therefore, to claim that space is infinite is also to claim that the aggregate of material things, upon which space depends for its existence, is also infinite. Leibniz wrote to Clarke: 'I don't say that matter and space are the same thing. I only say, there is no space, where there is no matter; and that space in itself is not an absolute reality.'[30] In other words, the world cannot be limited in space since the world is not *in* space (content to container), but space is *in* the world (predicate to subject, or relation to *relata*). In the antithesis the question of the finitude or infinitude of the world encompasses not only the magnitude of the material universe but space itself (in which we normally say the universe is situated) since the two are inextricably connected. The clear implication of the antithesis is that the idea of an empty space beyond the limits of the material universe is completely meaningless and absurd. The universe is a spatial *plenum*.

The proof of the antithesis assumes the idea of a finite world in absolute space and then raises a crucial question about the kind of relations that can possibly obtain between the material world and its infinite container. Then, it tries to show that whatever relations one may wish to suggest as obtaining between them they are impossible. I shall examine the three 'alleged' relations

[29] Leibniz's fifth letter, para. 55.
[30] Ibid. 62.

(or determinations in Kant's words) between the world and its container that Kant hints at.

(*a*) In the footnote to the proof of the antithesis Kant raises a question about the 'determination of the relation of the motion (or rest) of the world to infinite empty space' and indicates that from the point of view of the antithesis such a relation is completely unreal. 'It is a mere thought-entity.' Kant is clearly referring to the Newtonian idea that science can meaningfully talk and ask sensible questions about the absolute motion of the world as a whole relative to infinite space. And this is precisely what the antithesis denies in much the same way that Leibniz attacked Clarke's defence of the reality of the absolute motion of the world.[31]

(*b*) In the observation on the antithesis Kant raises the question of the reality of *vacua* in nature. The thesis necessarily entails the reality of absolutely empty space beyond the limits of the world *in* space. The antithesis

---

[31] The importance of this footnote in clarifying the proof of the antithesis seems to have completely escaped Norman Kemp Smith. He simply sees in it an attempt on the part of Kant to 'patch up' the deficiencies of the argument of the proof by reintroducing his theory of space as we find it expounded in the *Aesthetic*. Kemp Smith says: 'That Kant himself felt the inadequacy of this argument, when taken from the dogmatic standpoint, is indicated by the lengthy note which he has appended to it, and which develops his own Critical view of space as not a real independent object, but merely the form of external intuition.' (*Commentary*, p. 488.) Then he claims that this 'introduction of the opposed standpoint of the *Aesthetic*' lands Kant in certain perplexities, etc. . . . Martin drew attention to the importance of this footnote where, according to him, 'Kant alludes to the nonsensical consequences of such a representation of a finite world in an infinite space'. The following discussion of the argument of the antithesis is partly based on Martin's very brief treatment of these 'nonsensical consequences'. (*Kant's Metaphysics and Theory of Science*, pp. 49–50.)

categorically denies this implication and takes its false-hood as grounds for discrediting the premise from which it follows, namely, the finitude of the world. From the point of view of the antithesis space 'may be limited by appearances, but appearances cannot be limited *by an empty space* outside them'.[32] This is precisely the tactic which Leibniz followed in attacking Clarke's defence of the orthodox Newtonian position on this question.

(c) In the proof of the antithesis and the observation on it Kant, in effect, raises the question of the nature of *the spatial* relation that the world has to absolute space. Evidently he is referring, here, to the Newtonian con-tention that the world has an absolute determinate place (situation) in infinite space. All things are ordered in space (as Leibniz also holds) but in addition the aggre-gate of all things forms an entity (a *world*) which has a definite, and in principle specifiable, position relative to absolute immovable space. The antithesis means to deny this contention *in toto* as well as the premises underlying it.

Now since the world is an absolute whole beyond which there is no object of intuition, and therefore no correlate with which the world stands in relation, the relation of the world to empty space would be a relation of it to no object.[33]

These three points were sharply debated by Leibniz and Clarke in their famous correspondence. Leibniz rejected all three Newtonian claims on the grounds that they violated the law of sufficient reason as he inter-preted it. Concerning the first claim about the absolute motion of the universe he wrote in reply to Clarke:

To say that God can cause the whole universe to move

[32] Observation on the antithesis.
[33] The proof of the antithesis.

forward in a right line, or in any other line, without making otherwise any alteration in it; is another chimerical supposition. For, two states indiscernible from each other, are the same state; and consequently, 'tis a change without any change. . . . But God does nothing without reason; and 'tis impossible there should be any here.[34]

In the footnote to the proof of the antithesis Kant first raises the whole issue of the relations of the world to infinite space and then restates the essence of the Leibnizian objection to the idea of the absolute motion of the world as part of the support he can produce for the claim of the antithesis. Kant writes about the implications of the attempt to separate the world of appearances from space the following:

If we attempt to set one of these two factors outside the other, space outside all appearances, there arise all sorts of empty determinations of outer intuition,[35] which yet are not possible perceptions [i.e. they are indiscernible]. For example, a determination of the relation of the motion (or rest) of the world to infinite empty space is a determination which can never be perceived, and is therefore the predicate of a mere thought-entity.

Concerning the second claim Leibniz denied the reality of *vacua* on the grounds that since all parts of space are absolutely homogeneous God will have no sufficient reason to fill some parts of space rather than all of them. He wrote on this point:

I lay it down as a principle, that every perfection, which God could impart to things without derogating from their other perfections, has actually been imparted to them. Now let us fancy a space wholly empty. God could have placed

[34] Leibniz's fourth letter, para. 13.

[35] It should be recalled here that in critical terminology space is the form of outer intuition.

some matter in it, without derogating in any respect from all other things: therefore, he has actually placed some matter in that space: therefore, there is no space wholly empty: therefore, all is full.[36]

The third sort of empty determination, which arises on the assumption of a finite world, concerns the absolute position of the world in infinite space. Leibniz maintained that since all parts of space are wholly alike there can be no sufficient reason why the world should be located in one region of empty space rather than any other. He says:

Now from hence it follows, (supposing space to be something in itself, besides the order of bodies among themselves) that 'tis impossible there should be a reason, why God, preserving the same situations of bodies among themselves, should have placed them in space after one certain particular manner, and not otherwise; why everything was not placed the quite contrary way, for instance, by changing East into West.[37]

This Leibnizian position reappears in the arguments for the antithesis but without its theological complications. In the observation on the antithesis it is stated in the following words:

. . . if the world has limits in time and space, the infinite void must determine the magnitude in which actual things are to exist. . . .

While in the section of the proof of the antithesis pertaining to time Kant states the same principle by saying that no part of empty time 'possesses, as compared with any other, a distinguishing condition of existence rather than of non-existence'. In another comment on the proof

[36] Postscript to Leibniz's fourth letter.
[37] Leibniz's third letter, para. 5.

of the antithesis Kant reproduces the essential point of the Leibnizian argument with complete accuracy. He wrote:

In that argument [proof of the antithesis] we regarded the sensible world, in accordance with the common and dogmatic view, as a thing given in itself, in its totality . . . and we asserted that unless it occupies all time and all places, it cannot have any determinate position whatsoever in them . . . in (that) dogmatic proof we inferred the actual infinity of the world. (A521–B549 n.)

In other words, from the point of view of the antithesis, if the world is finite then space cannot be homogeneous since that part of space in which the world is actually located must possess 'a distinguishing condition of existence rather than non-existence' as compared to the other regions of empty space. Presumably this distinguishing characteristic acts as God's sufficient reason for placing the universe in precisely that region of space and no other. Now, if the defenders of the thesis insist on the homogeneity of space, as they have to, then either the world is infinite or a violation of the law of sufficient reason would have occurred. Since the latter can never occur it follows that the world fills all spaces. It is infinite.

It should be evident from this discussion that the strength of Kant's proof for the antithesis derives from the general Leibnizian position and arguments against the claims of Newton and Clarke which permeate the whole of the first antinomy. Furthermore, it should be evident that the law of sufficient reason, as interpreted by Leibniz, is absolutely essential for supporting and maintaining the explicit and implicit claims of the antithesis. The argument in favour of the antithesis is, therefore, not as inadequate and shabby as has often been thought by those who have examined the first antinomy in isolation from its supporting historical background.

The claims of the thesis and antithesis represent for Kant a good example of an irresolvable metaphysical dispute between two dogmatic philosophers. Each disputant argues for his thesis in an *a priori* fashion and is unwilling to specify the kind of empirical conditions that should prevail in order to render his point of view scientifically acceptable. Contrary to Weldon's interpretation the impasse that the antinomy presents is not, really, between empiricism and rational cosmology.[38] It is between the dogmatic 'mathematical students of nature' (i.e. the Newtonians who went beyond the legitimate conclusions of their science to construct a 'metaphysics' for which they dogmatically claimed a privileged status in view of the prestige of their achievements in the study of certain natural phenomena) and the dogmatic 'metaphysical students of nature' (i.e. Leibniz and Wolff) as Kant calls them[39] (B56–A40). The point is that both are making dogmatic claims that can be justified neither empirically nor logically. Thus, Kant characterizes the antinomy as:

   . . . the conflict of the doctrines of seemingly dogmatic

[38] It would be instructive to note in this connection that, contrary to the prevalent view, Windleband points out that Kant did not mean to characterize by 'dogmatic' only rationalism but also empiricism especially in its pre-Humean versions. (*A History of Philosophy*, Harper Torch Books, 1958, vol. ii, p. 537 n.) Certainly the opposite of the 'dogmatic temperament' is not simply 'empiricism' but 'the critical spirit'. Empiricism can be as dogmatic as rationalism in asserting its claim and principles.

[39] In the *Dissertation* he refers to them as 'the English' and 'our countrymen' respectively (61–2).

knowledge (*thesis cum antithesi*) in which no one assertion can establish superiority over another. (A421)[40]

In effect each of the thesis and the antithesis forms a self-contained system of ideas which cohere with each other and are, as such, sound and compelling. However, no rational choice between them is possible on *a priori* grounds. Kant makes this point clear when he writes:

> ... there arises an unexpected conflict which never can be removed in the common dogmatic way; because the thesis, as well as the antithesis, can be shown by equally clear, evident, and irresistible proofs. . . .[41]

No rational choice between them is possible on *a posteriori* grounds either; as Kant elaborates:

> For how can we make out by experience whether the world is from eternity or had a beginning . . .? such concepts cannot be given in any experience, however extensive, and consequently the falsehood either of the affirmative or of the negative proposition can not be discovered by this touchstone.[42]

Here the question arises as to why we cannot settle the conflict by means of *a priori* considerations? In this connection we should note that, according to Kant, in a conflict which is genuinely antinomial, the assertion and counter assertion follow from 'universally acknowledged principles'.[43] I think that the principles he has in mind ('which every dogmatic metaphysics must necessarily recognize . . .')[44] are the law of contradiction and the law of sufficient reason. It is the latter that really concerns us in the context of the first antinomy since we are deal-

[40] See also the first paragraph of his observation on the thesis.
[41] *Prolegomena*, pp. 339–40.
[42] [43] Ibid., pp. 340–1.
[44] Ibid., pp. 378–9.

ing with the realm of actuality and not with the realm of possibility only.

It is part of the procedure of the two dogmatic disputants represented in the antinomy first, to acknowledge verbally the universal principle of sufficient reason[45] and then to interpret it in incompatible ways that make it possible for each one of them to derive his assertion (or counter assertion) from the principle itself. Then there are really two principles of sufficient reason involved: one worked out to suit the claims of the thesis while the other is made for the claims of the antithesis. Each disputant is, in a sense, right in claiming that his assertion follows coherently from the principle of sufficient reason. Thus the conflict cannot be resolved in an *a priori* fashion because there is really no single higher rational principle common to both disputants that can serve as our *a priori* touchstone to favour one claim as against the other. Each disputant will simply and dogmatically repeat that his interpretation of this ultimate principle is the 'right' one.

This situation is clearly demonstrated in the controversy between Leibniz and Clarke to which we had occasion to refer. Both philosophers claimed to accept the law of sufficient reason as an ultimate principle determining the nature of actuality. They agreed to the principle 'that nothing happens without a sufficient reason, why it should be so, rather than otherwise'.[46] But, then, when they utilized the principle to produce proofs for their incompatible positions about the finitude

---

[45] For instance, Clarke says: 'that in general there is a sufficient reason why everything is, which is; is undoubtedly true, and agreed on all hands'. (Fifth letter, paras. 124–30.)

[46] Leibniz's third letter, para. 2.

of the world each one of them interpreted it in such a manner as to make his thesis consistent with the principle. Given Leibniz's interpretation of the principle it logically follows that the world cannot have a limit in space. Given Clarke's interpretation of the same principle it follows that the world has to have a limit in space. And each of the two philosophers dogmatically insisted that his own interpretation is the 'true' or 'correct' one but without being able to produce any arguments or justifications in support of his claim. For Clarke God's will is enough of a sufficient reason for the world to occupy the magnitude of infinite space that it actually does and the fact that the universe exists shows that the utter homogeneity of the points of space presents no hindrance to God's exercise of choice. On the other hand, according to Leibniz God cannot exercise His choice unless there is a sufficient reason for Him to do so external to His will. Therefore, if the points of space are truly homogeneous then God would not have chosen a situation for the world in empty space.

I have had occasion to show that all these ideas are reflected in the first antinomy of pure reason and that the proof of the antithesis is really based on Leibniz's version of the law of sufficient reason. Now I would like to use this opportunity to demonstrate the connection between the thesis and its proof, on the one hand, and a second version of the principle of sufficient reason, on the other.

To elucidate this connection we should note that the thesis really asserts two points: the first consists in the negation of the claim of the antithesis to the effect that the world has no limits in space; and the second is the positive claim that the world does have a limit in a specific portion (or magnitude) of space and no other. In other words, the thesis is not only asserting that if the world exists then it must have a limit in space, but is also asserting that the world does exist and as such it has a certain definite position relative to infinite space and no other.

Now, the logico-mathematical argument of the proof of the thesis is meant to support the first point. The second point cannot be established without appeal to the law of sufficient reason. That is if the Newtonian is to give 'satisfactory' answers (as Clarke felt that he ought to) to such questions as why did the world come to exist in space rather than remain an ideal possibility? and why does it exist in this particular region of homogeneous space and no other? then he has to appeal to the law of sufficient reason or to some version of it. And this is exactly what Clarke did.[47] It would have been almost heroic for Clarke to declare that these questions are, in principle, unanswerable. This is what Kant did when he declared that in so far as these and similar questions deal with matters that are altogether beyond appearances (i.e. experience), they cannot be sensibly answered. Furthermore, Kant made it quite clear that the Newtonian idea of an actual finite universe (in infinite space

[47] The following typical statement of Clarke represents his understanding and application of the principle of sufficient reason: "'Tis very true, that nothing is, without a sufficient reason why it is, and why it is thus rather than otherwise. And therefore, where there is no cause, there can be no effect. But this sufficient reason is oft-times no other, than the mere will of God. For instance: why this particular system of matter, should be created in one particular place, and that in another particular place; when, (all place being absolutely indifferent to all matter), it would have been exactly the same thing vice versa, supposing the two systems (or the particles) of matter to be alike; there can be no other reason, but the mere will of God.' (Clarke's second letter, para. 1. See also his third reply, para. 2.)

and time) cannot be maintained without appeal to some principle which expresses the 'distinguishing condition of existence' favouring its actual existence over its non-existence. Without such an appeal all that the thesis can assert is that if the world were actual then it must have a definite position (for example) relative to infinite space and cannot proceed to hold, as it does, that the world actually does have a definite and, in principle, determinable position in absolute space.

We may say, then, that the thesis presupposes the idea that since the world is actual the homogeneity of the points of space could not have constituted a hindrance to the ruling of the principle of sufficient reason which precludes any location of the world in infinite space other than the actual one which obtains. Clarke wrote on this point: 'The uniformity of all the parts of space, is no argument against God's acting in any part, after what manner he pleases.'[48] This will hold true provided we do not insist that 'this distinguishing condition' defining the sufficient reason for the world to be in that particular location and no other should be inherent in the parts (points) of space themselves. In other words, all will go well provided we interpret the law of sufficient reason in a manner that will make it consistent with the claim of the thesis concerning the finitude of the world and thus render the antithesis inconsistent with this version of the principle.

We might look at the matter from a slightly different angle. Both the thesis and antithesis agree that if the principle of sufficient reason is suspended then the idea of the finitude of the world is as consistent with the law of contradiction as the idea of its infinitude. In this sense

[48] Fourth letter, para. 18.

the law of contradiction cannot serve as our *a priori* touchstone for settling the conflict between the two opposed claims. Since, in addition to the law of contradiction, actuality is also subject to the law of sufficient reason then both thesis and antithesis have to appeal to it for the vindication of their claims. But the net result is an impasse and a stalemate since when the thesis appeals to the law, it implicitly interprets it in such a manner as to make a spatial limit of the world a necessary result of the law as explained above. We already saw that the spatial finitude of the world is part of the Newtonian definition of a *world*. Now we add that its finitude is, therefore, a necessary condition of its existence. On the other hand, when the antithesis appeals to the principle of sufficient reason it interprets it in a fashion that makes the infinitude of the world a necessary consequence of the principle.

The moral of the entire episode of the antinomy for the critical philosophy as a whole is threefold:

(*a*) That such rational principles, as the law of sufficient reason, are purely formal principles from which nothing can be inferred about the nature of actuality. This is certainly a central doctrine in the critical philosophy.

(*b*) That all such metaphysical disputes are futile and senseless in so far as their assertions and counter assertions claim to give us synthetic (and also *a priori*) knowledge about the nature of things as they are in themselves, viz. about objects that are in principle beyond all the possible bounds of experience. Kant expresses his attitude towards this matter in the following words:

We may blunder in various ways in metaphysics without

any fear of being detected in falsehood. If we but avoid self contradiction, which in synthetical though purely fictitious propositions is quite possible, then whenever the concepts which we connect are mere ideas that cannot be given (with respect to their whole content) in experience, we cannot be refuted by experience.[49]

(c) That space is transcendentally ideal. Kant utilizes the difficulties inherent in the first antinomy to strengthen the conception of space expounded in the *Aesthetic*. Space is neither dependent upon existents nor is it a thing existing independently of all experience but an appearance that has no reality apart from a representing agent and in this fact lies its empirical reality as well as its transcendental ideality. Kant writes on this point:

From this antinomy we can, however, obtain, not indeed a dogmatic, but a critical and doctrinal advantage. It affords indirect proof of the transcendental ideality of appearances—a proof which ought to convince any who may not be satisfied by the direct proof given in the Transcendental Aesthetic. This proof would consist in the following dilemma. If the world is a whole existing in itself, it is either finite or infinite. But both alternatives are false (as shown in the proofs of the antithesis and thesis respectively). It is therefore also false that the world (the sum of all appearances) is a whole existing in itself. From this it then follows that appearances in general are nothing outside our representations—which is just what is meant by their transcendental ideality. (A506-7, B534-5)

There is one comment to be made about this citation: strictly speaking Kant cannot infer that the two sides of the antinomy are false since he made it quite clear that

49 *Prolegomena*, pp. 340-1.

neither pure reason nor experience can refute either of the two claims. It would be more accurate to infer that neither thesis nor antithesis is really making a genuine claim about appearances and that, therefore, the conflict between them is irrelevant since human knowledge is limited to appearances.

The explanation of Kant's inference that both sides of the antinomy are false is to be found in the *Prolegomena*[50] where he presents us with a purely 'logical' interpretation of the nature of the antinomial conflict. According to this account the two contradictory propositions forming the antinomy are both false because 'the concept on which each is founded is self-contradictory'.[51] Kant illustrates his meaning by taking the following two contradictory propositions each of which is founded on a self contradictory concept: 'a square circle is round' and 'a square circle is not round'. Both of these propositions are false for '. . . it is false that the circle is round because it is quadrangular, and it is likewise false that it is not round, that is, angular, because it is a circle'.[52]

Kant, then, proceeds to identify the self-contradictory concept on which each of the thesis and antithesis of the antinomy is supposedly founded. This he finds in the contradiction which exists between 'the concept of a world of sense' and the concept of a world existing in itself apart from any experience: i.e. 'the concept of an absolutely existing world of sense'. Kant explains himself in the following words:

Hence it follows that, as the concept of an absolutely existing world of sense is self-contradictory, the solution of

[50] Part III, sections 52b, 52c, 53.
[51] [52] *Prolegomena*, pp. 341–2.

the problem concerning its magnitude (i.e. the first antinomy) whether attempted affirmatively or negatively, is always false.[53]

By following Kant's example of the square circle we can reconstruct his meaning in the following manner: Thesis: 'The absolutely existing world of sense' is finite. Antithesis: 'The absolutely existing world of sense' is infinite. The thesis is false because it says that the world exists 'absolutely' (i.e. as it is in itself apart from all experience) and that this sort of world is finite. The antithesis is false because it says that the world exists as a world of sense and is also infinite.

This is probably the most artificial, abstract, and unsatisfactory interpretation of the nature of the antinomial conflict that we can find in Kant's writings. This interpretation of the antinomy on the analogy of Kant's initial example of the square circle will simply not work without the implicit introduction of an arbitrary premise external to the argument. For Kant is capable of asserting the falsity of the proposition 'the square circle is round' because the 'squareness' in the subject concept excludes automatically the 'roundness' contained in the

---

[53] *Prolegomena*, pp. 342–3. Kant explains the source of this contradiction at some length. He writes: 'I must not say of what I think in time or in space, that in itself, and independent of these my thoughts, it exists in space and in time, for in that case I should contradict myself; because space and time, together with the appearances in them, are nothing existing in themselves and outside of my representations, but are themselves only modes of representation, and it is palpably contradictory to say that a mere mode of representation exists without our representation. Objects of the senses therefore exist only in experience, whereas to give them a self-subsisting existence apart from experience or before it is merely to represent to ourselves that experience actually exists apart from experience or before it.' (*Prolegomena*, pp. 341–2.)

predicate concept. But he cannot analogously assert the falsity of the proposition 'the absolutely existing world of sense is finite' because the 'absoluteness of the world' does not automatically and prima facie exclude the 'finitude' asserted in the predicate concept. In other words, without implicitly accepting the extraneous premise that 'the world as it exists absolutely cannot be finite', Kant's analogy and argument will not work. In fact all such assertions about the 'absolute world as it exists in itself' are pointless from the point of view of synthetic knowledge as understood by the critical philosophy.

Similar considerations apply to the case of the antithesis. Kant asserts the falsity of the proposition 'the square circle is not round' on the grounds that the 'circularity' in the subject concept automatically excludes the denial of 'roundness' contained in the predicate concept. But this can in no wise apply to the proposition of the antithesis, viz. that 'the absolutely existing world of sense is infinite'. For the 'world of sense' in the subject does not automatically and prima facie exclude the 'infinitude' asserted in the predicate.

Furthermore, Kant ignores completely the fact that his initial example about the square circle is capable of the opposite interpretation. The two propositions: 'a square circle is round' and 'a square circle is not round' may be considered also as both true. The former is true because it is true that the circle is round. The latter is true because the circle in question is also 'not round' for it is quadrangular as the subject concept asserts. Similarly, even if we grant Kant that the thesis and the antithesis of the antinomy are each founded on a self-contradictory concept we may still conclude that both propositions are true (or may be true) rather than that both are definitely false.

This unsatisfactory treatment and interpretation, on the part of Kant, of the nature of the antinomial conflict brings us back to the point I made earlier to the effect that, strictly speaking, Kant may not infer that the two sides of the antinomy are both false since neither pure reason nor experience can refute either of the two claims. This conclusion remains closer to the over-all spirit of the 'critical' treatment of the antinomy than any other.

There remains two points in the observations on the antinomy about which I have not had the chance to comment:

1. In the observation on the thesis Kant insists on rejecting 'the defective concept of the infinitude of a given magnitude' in favour of the transcendental concept of infinitude as explained throughout the discussion. The defective concept is not 'adequate to what we mean by an infinite whole'. For it holds that 'a magnitude is infinite if a greater than itself, as determined by the multiplicity of given units which it contains, is not possible'. This defective notion of the mathematical infinite is explained in the *Dissertation* as 'an aggregate (of some assignable unit) than which a greater is impossible' (37 n.). Then Kant criticizes the holders of this defective view by saying that 'they here define not "infinite" but "maximum"...'.

It seems that Kant is here rejecting the Cartesian idea of the infinite. Descartes drew a sharp distinction between the infinite and the indefinite. For him an infinite whole is one in which all degrees of augmentation are already given (the 'maximum' mentioned by Kant). Descartes explained himself on this matter as follows:

Je ne me sers jamais du mot d'infini pour signifier seule-
ment n'avoir point de fin, ce qui est négatif et à quoi j'ai
appliqué le mot d'indéfini, mais pour signifier une chose
réelle, qui est incomparablement plus grande que toutes
celles qui ont quelque fin.[54]

Accordingly, the name infinite is reserved, according to
Descartes, to God alone for only in Him do we discover
no limits of any kind.

2. In the observation on the antithesis Kant recog-
nizes the attempt on the part of some to argue that it is
quite possible to think of the world as limited in space
and time without 'our having to make the impossible
assumption of an absolute time prior to the beginning
of the world, or of an absolute space extending beyond
the real world'. Again, it seems that Kant is referring to
the Cartesians and their conception of the world-whole
as an 'indefinite' but not 'infinite' plenum (in the strict
Cartesian sense of these terms). According to this conception the world is not infinite and at the same time it
is not bounded by empty space and time for it is a
plenum.

Kant considers this attempt to slip through the two
sides of the antinomy to be a plain evasion of the issue.
Therefore, he devotes a good portion of the observation
on the antithesis to finding fault with this procedure,
on account of its identification of the sensible world
(matter) with the conditions of its existence (space and
time).

[54] Letter to Clerselier, Adam & Tannery, vol. v, p. 356.

APPENDIX

TIME IN THE FIRST ANTINOMY

THE thesis of the antinomy asserts the finitude of the world-series in time and not of the temporal container itself. In other words, according to the thesis the world-series has a first limiting term which can be (theoretically) unambiguously correlated with one of the flowing moments of absolute time. That moment will give us the date at which the world-process began.

Following is a summary and explanation of the proof of the thesis:

1. Assume that the series of 'successive states' (forming the world) in time has no beginning.

2. This means that up to any chosen moment of receptacle-time an infinite number of the members of the series would have completely elapsed (or would be completed).

3. However, 'the infinity of a series consists in the fact that it can never be completed through successive synthesis'.

4. It follows that, up to any moment of time, it is impossible for an infinite number of successive states to have completely elapsed.

5. Therefore, a beginning of the series of successive states forming the world is a necessary condition of the existence of the series. Otherwise we would have to put up with the absurd idea of a series which is infinite and at the same time complete.

Now given Kant's definition of a *world* we can see that the idea underlying the whole proof is that the members of the series of successive states (the world) can be fully run through by 'successive synthesis' in 'a finite and assignable' number of the moments of receptacle-time. This is what constitutes the temporal finitude of the material universe. Accordingly we can reformulate the argument of the proof of the thesis in the following manner:

(a) On the supposition that the series of successive states in receptacle-time is infinite, then it can never form a *world*-series because in that case the 'successive synthesis' of its members can never be completed.

(b) But the series of successive states does form a *world*, i.e. the successive synthesis of its members can be, in principle, completed in 'a finite and assignable' number of the moments of time.

(c) Therefore, the series is finite (it has a beginning).

This is essentially the Newtonian conception of the world as a material process spread over a definite and, in principle, specifiable number of the 'equably' flowing instants of absolute time. Laplace's intellect would have no problems informing us about the exact number of these moments.

Strawson criticizes Kant's treatment of the subject in this antinomy on the grounds that he raises the question about the beginning of the world as a whole as if it were a question about the beginning of any ordinary process in the world.[1] According to Strawson such a question is meaningful only if an 'external temporal framework' is specified (either explicitly or implicitly) in terms of which we can account for the process's beginning. Strawson claims that Kant's argument fails to take this fact into consideration. However, it should be clear from the above treatment that, contrary to what Strawson suggests, Kant did not raise the question about the beginning of the world *in vacuo*, i.e. without reference to a 'temporal framework' external to the world-process itself. The temporal framework is obviously the uniform (equable) flow of the moments of receptacle

[1] *The Bounds of Sense*, p. 178.

time. It is in terms of this framework that the beginning of all processes, including that of our world, are determined. This is often stated by Clarke in his letters to Leibniz. He writes:

The order of things succeeding each other in time, is not time itself: for they may succeed each other faster or slower in the same order of succession, but not in the same time. If no creatures existed, yet the ubiquity of God, and the continuance of his existence, would make space and duration to be exactly the same as they are now.[2]

According to the antithesis the world is infinite as regards time (it has no beginning). The proof of the antithesis may be summarized and explained in the following points:

1. Assume that the world-series began at a certain specifiable moment (M) of the flowing and homogeneous moments of infinite receptacle-time.

2. This means that the moments of time preceding M are empty.

3. It also means that for the world-series to begin at that moment and no other M possessed some characteristic which distinguished it from the other moments of time. This is so because 'no coming to be of a thing is possible in an empty time' unless 'such a time possesses, as compared with any other, a distinguishing condition of existence rather than non-existence'.

4. But according to the defenders of the thesis the moments of receptacle-time are absolutely homogeneous. M could not possibly have a characteristic ('a condition of existence') distinguishing it from all the other moments of time.

5. Therefore, the world-series never began at M or any other similar moment of time. It has always been there.

6. Hence the conclusion of the proof: 'In the world many series of things can, indeed, begin; but the world itself can

[2] Fourth letter, para. 41.

not have a beginning, and is therefore infinite in respect of past time.'

The antithesis and its proof harmonize very well with the Leibnizian idea which rejects the distinction between the absolute temporal receptacle on the one hand and the world-process enduring in it on the other. Leibniz wrote to Clarke '. . . instants considered without the things, are nothing at all . . .'[3] and 'If there were no creatures there would be neither time nor place . . .'.[4]

We should note also that the proof proceeds by assuming the idea of a finite world in the absolute temporal receptacle and then raising a basic question about the relationship of the first state of the world-series to the homogeneous moments of time. The question is why should the first state of the world series occur at any one moment of time rather than another when all moments are wholly alike? It is the problem of the principle of sufficient reason all over again. Furthermore, this is exactly the point which Leibniz kept urging against Clarke and the Newtonians. Leibniz wanted to put the opposition before a difficult choice: granted the actuality of the world-series, then either the moments of time are not all homogeneous or the world never began at any one of these moments. And Clarke always escaped the choice by appealing to the will of God.

[3] Third letter, para. 6.
[4] Fifth letter, para. 106.

THE SECOND ANTINOMY

THE second antinomy of pure reason deals essentially with the problem of the infinite divisibility of substance. Naturally it also touches upon some subsidiary issues which arise from that problem. Norman Kemp Smith specifies the precise subject of the antinomy as follows: 'The substance referred to, though never itself mentioned by name, is extended matter.'[1] However, this statement is not quite accurate, on two counts. In the first place, in the observation on the antithesis, Kant explicitly mentions that the 'substance' of the antinomy includes matter, bodies, and the objects of outer intuition. In the second place the antinomy is not exclusively devoted to the exposition of conflicting views on the constitution and structure of extended material substance. It also takes into account the traditional notion of mental substance. This aspect of the antinomy is generally overlooked and Smith ignores it completely.

I think that Kant constructs the antinomy in terms of 'substance' in general because he meant to cover the two traditionally recognized types of substance, viz. the material and the mental. It would be worth while to mention here that material and mental substance have been treated conjointly by modern philosophers. Thus most thinkers who adopted (or expounded) a specific view about the nature of material substance (or mental substance) felt constrained to adopt, either explicitly or

[1] *Commentary*, p. 489.

implicitly, a corresponding theory about the nature of mental substance (or material substance) which they felt is consistent with their first theory. This applies even to 'reductionistic' theories which consistently regard the nature of one of these substances as ultimately derivative from the other. Accordingly, arguments in the antinomy dealing specifically with material substance have their immediate and consistent implications for the analogous and complementary position about mental substance. However, these implications are only occasionally stated in the antinomy. The very same consideration applies to any argument in the antinomy dealing specifically with the nature of mental substance. Wherever necessary, the texts of the antinomy indicate which of the two substances is specifically being considered. Consequently, in this conflict of pure reason all the arguments pertaining to a certain specific conception of the nature of material and mental substance hold together and reinforce each other to form a definite systematic position about the nature of 'substance' in general. The antinomy is made up of two such systematic positions which are in conflict with each other on account of their contradictory claims about the supposed nature of substance.

In so far as the second antinomy deals with the constitution and structure of material substance then it obviously forms the complementing counterpart of the first antinomy. In so far as it also deals with the nature of mental substance it simply goes somewhat beyond the basic concerns of the first antinomy. The first antinomy dealt with the question of whether the process of synthesizing material substance into greater and greater wholes is finite or infinite. The second antinomy deals with the question of whether the process of analysing (subdividing) material substances into smaller and smaller parts is finite or infinite. That Kant did see the first two antinomies in this sort of conjunction is made clear in the *Dissertation* of 1770 where he says: 'Further, if a complex of substances be given, whether through the testimony of the senses or in any other way, it is easy to see, by an argument based on intellectual grounds, that simples, and a world, are also given' (38).

It is obvious that the issues involved in the first two antinomies can be formulated either in the language appropriate to spatial extension or the language appropriate to temporal extension. The first antinomy explicitly formulates the problem of the finitude or infinitude of the material universe in both manners. The second antinomy is dominated by the language and terminology appropriate to space. It should be clear that all the arguments in the second antinomy can be restated in the terminology of temporal duration without altering their signification in the least. Indeed, P. F. Strawson's discussion of the second antinomy in *The Bounds of Sense* abandons the cumbersome terminology of space in favour of the more convenient terminology of temporal idioms.

II

The thesis of the second antinomy states:

Every composite substance in the world is made up of simple parts, and nothing anywhere exists save the simple or what is composed of the simple.

The proof of the thesis may be summarized in the following points without departing too much from Kant's own terminology and manner of statement:

48

1. Assume that: 'Composite substances are not made up of simple parts.'

2. Now let us perform the following hypothetical experiment: 'If all composition be then removed in thought, no composite part, and (since we admit no simple parts) also no simple part, that is to say, nothing at all, will remain, and accordingly no substance will be given.'

3. Hence, either (a) 'It is impossible to remove in thought all composition', or (b) nothing exists at all, i.e. no substance(s) is given whatsoever.

4. Concerning (a) it is not impossible to remove in thought all composition because 'composition, as applied to substances, is only an accidental relation in independence of which they (the simple substances) must still persist as self-subsistent beings'. While (b) contradicts our initial assumption and violates the definition of substance which says that substances 'persist as self-subsistent beings'. Therefore some substance is actually given.

5. 'It follows, as an immediate consequence, that the things in the world are all, without exception, simple beings; that composition is merely an external state of these beings; . . .'

The text of the proof ends with the explanation that the argument for the thesis makes no appeal to empirical considerations. It admits that in actual fact 'we can never so isolate these elementary substances as to take them out of this state of composition . . .'. But still, from the point of view of the advocates of the thesis, 'reason must think them (the simples) as the primary subjects of all composition, and therefore, as simple beings, prior to all composition'.

At this point it becomes necessary to emphasize that the thesis and its proof deal with the constitution of space-occupying material substance and not with the nature and structure of the spatial container itself. I dwell on this initially obvious point because Kant went out of his way to define precisely what the thesis is about and commentators have often gone out of their way to ignore Kant's effort in this direction. Consequently, some criticisms of Kant's handling of the thesis and its proof remain muddled precisely because they ignore the distinction drawn by Kant between the composite material substance and its spatial container.

That the thesis deals only with the composition of space-occupying matter is made unambiguously clear in the proof of the antithesis (which assumes the thesis and then unpacks its implications in order to refute it). That Kant did not intend the argument of the proof of the thesis to encompass such things as space, time, and motion is made very clear in the observation on the thesis. Kant argues there that although space, time, and motion may be considered 'composites' in a certain general sense, they are not to be so regarded in the sense in which material substances are composite. According to the thesis space-occupying matter is composite in the strict and literal sense of being composed out of simple irreducible parts. Space and time, as we had occasion to note more than once, are analytic totalities. Their 'parts are possible only in the whole' and 'not the whole through the parts'.[2] Kant says in this connection:

Space and time do not, therefore, consist of simple parts. What belongs only to the state of a substance, even though it has a magnitude, e.g. alteration, does not consist of the

[2] Observation on the thesis.

simple; that is to say, a certain degree of alteration does not come about through the accretion of many simple altera-tions.[3]

Then he sounds his warning against abusing the argu-ment for the thesis by extending its application beyond the legitimate limits:

Thus the proof of the necessity of the simple, as the constitutive parts of the substantially composite, can easily be upset (and therewith the thesis as a whole), if it be extended too far and in the absence of a limiting qualifica-tion be made to apply to everything composite—as has frequently happened.[4]

Norman Kemp Smith makes the following comment on this citation from Kant's observation: 'In the *Observation* on this thesis Kant shows consciousness of the defects of his argument. It does not apply to space, time, or change.'[5] Far from being a defect in the argu-ment the limitations imposed by Kant upon the ap-plicability of the proof of the thesis constitute its very point of strength. Among other things, the limitations make the proof consistent with Kant's general view that space and time are analytic totalities and not synthetic totalities. Smith could have come to such a conclusion as this (in spite of Kant's clear warnings) only as a result of his failure to distinguish clearly between the material substance which is the subject of the claims of the thesis and its proof, on the one hand, and the space contain-ing it, on the other. About the latter the thesis pretends to make no special claims. Failure to draw this distinc-tion has led Kemp Smith to want to apply what can be

[3][4] Observation on the thesis.
[5] *Commentary*, pp. 489–90.

51

properly said only of matter in space to space itself. And, as a result, the fact that Kant does not favour such an extended application of the argument appears as a defect to Smith.

The reader will remember that Weldon and Kemp Smith have identified the theses of the antinomies with the claims of the rationalist cosmologists of the Continent and in particular with the dogmatic rationalism of the Leibnizians.[6] However, the fact is that the thesis of the second antinomy is a fairly representative statement of the dogmatic metaphysics which grew around the Newtonian science of motion. The thesis is predicated upon the atomic theory of matter as expounded by Newton and such Newtonians as Clarke, Euler, and the Kant of 1768 to 1770.[7] According to this theory matter is constituted out of indivisible particles which change their positions in absolute space relatively to each other and to immovable space itself. This fact explains Kant's legitimate insistence that the proof of the thesis would be invalidated if extended to apply to space and time themselves. On the Newtonian theory of the world there is no necessary connection or organic interdependence between the nature of absolute infinite space on the one hand and the finite amount of matter placed in it on the other, beyond the external and accidental relation of container to contained.

One of the central ideas in the proof of the thesis is the claim that if we think away all composition in a given

[6] E. Caird also traces the proof of the thesis to Leibnizian philosophical views. (*The Critical Philosophy of Kant*, vol. ii, p. 46.)

[7] In 1768 Kant published a short essay 'On the First Ground of the Distinction of Regions in Space', which marked his short-lived conversion to a fully Newtonian point of view concerning the nature of space, time, matter, etc. . . .

composite material substance, without at the same time assuming the reality of simple atomic particles, then nothing would remain at all in the world. In other terms, no matter would then be present in container space. But since it is admitted on all hands that matter is present in the world it follows that . . . etc. Now the idea of this argument comes directly from the Leibniz–Clarke correspondence. It is a somewhat more elaborate statement of one of Clarke's main arguments against Leibniz and in favour of the strictly atomistic theory of matter. Clarke formulates the argument in the following two ways:

But if there be no such perfectly solid particles, then there is no matter at all in the universe.

Again:

If therefore, carrying on the division *in infinitum*, you never arrive at parts perfectly solid and without pores; it will follow that all bodies consist of pores only, without any matter at all: which is a manifest absurdity.[8]

Similarly, the idea of composition (compositeness) as interpreted in the thesis and its proof is in full accordance with the Newtonian conception of the nature of the composite in space and of the sort of relations which hold among its parts. Accordingly compositeness is seen in the thesis and its proof as a purely accidental and external relation among simple substantial elements. Kant states in the proof:

Composition, as applied to substances, is only an accidental relation in independence of which they must still persist as self-subsistent beings.

The elements which enter into the accidental relationship of composition are simple in the sense that

[8] N.B. to Clarke's fourth letter.

each one of them is 'a part which is not a whole'.[9] They are substantial in the sense that they are 'self-subsisting things'.[10] And they are atomic in the sense that they are initially 'given as *separate*'[11] in space regardless of any possible relations of composition that they may enter into later on. In the observation on the thesis Kant calls these simple elements atoms (he uses the term as a masculine: *den atomus*) in order to distinguish them from simple immaterial substances (*monas*) such as 'the simple which is *immediately* given as simple substance (e.g. in self consciousness) . . .'.[12]

This notion of 'composition out of parts' contained in the thesis and its proof is, in effect, a restatement of Clarke's views of the matter. Clarke briefly characterizes the fundamental nature of simple parts as 'separable, compounded, ununited, independent on and moveable from, each other'.[13] In other words, the thesis and what belongs to it is no more than an embodiment of the metaphysics of 'simple location' to use Whitehead's famous phrase.

Accordingly, we can see that the claims of the thesis and its proof reduce themselves to the following general position: that composite substances in the world should be made up of simple parts is a necessary condition of their actual existence since according to Newtonian science 'being made up of simple elements' is part of the 'true' and 'real' definition of 'compositeness'. This definition of 'compositeness' is not a 'mere definition' or an arbitrary account of the concept similar to those traditionally produced by metaphysicians and specula-

---

[9] *Dissertation*, sec. i, para. 1.
[10] Observation on the thesis.
[11] [12] Observation on the thesis (Kant's italics).
[13] Fourth letter, paras. 11 and 12.

tors. It is the 'scientific' account of the nature of compositeness. It is the 'true' account of the constitution of physical reality as is attested by the immense success of Newton's mathematical system of 'natural philosophy'. If we were to resort again to Laplace's superhuman intellect, we would say, then, that according to the thesis when such an intellect thinks away all 'accidental' and 'external' relations of composition in material substance its instantaneous gaze would necessarily fall upon a finite number of simple independent and self-subsistent atomic particles spread throughout a portion of absolute space. Since the human mind is not equipped with such powers of perception it resorts to abstract processes of analysing and subdividing the composite in order to represent to itself the constituent simple parts. In other words, the thesis works to draw the implications of the Newtonian definition of 'compositeness' with the implicit assumption that such are true and accurate descriptions of the nature of the material universe.

We can see from this that a certain criticism urged by Norman Kemp Smith against Kant's acceptance of the above definition of the 'compositeness of matter' is strictly irrelevant and betrays a basic misunderstanding as to what the thesis is all about. Kemp Smith maintains in this criticism that Kant makes no attempt to prove that the compositeness of matter can be defined in the above fashion.[14] However, Kant does not need to produce any such 'proof' considering that the above definition of the compositeness of matter is absolutely fundamental in the Newtonian conception of physical reality. The thesis and its proof state this conception

[14] *Commentary*, p. 489.

but, strictly speaking, they are not called upon to defend it. In addition, the definition has had in its favour the entire weight of authority carried by the greatest single scientific achievement of modern times: the Newtonian science of motion. Would it be reasonable to expect Kant to have given a better proof than this? Furthermore, Kant's objective in constructing the antinomy is not the 'improvement' of any of the two dogmatic metaphysical and conflicting sides but the exposition of the illusoriness of all such metaphysical disputes and dogmatic conflicts.

Here we should carefully note that the thesis starts by denying the claim of the antithesis to the effect that 'no composite thing in the world is made up of simple parts' and then proceeds to make the positive assertion that all material substances are in fact made up of simple parts. In other words, the thesis is not only asserting that (a) if composite substances exist then they must be composed of simple parts (or if mental substances exist then they must be absolutely simple), but is also asserting that (b) such substances do actually exist and as such they are in fact made up of simple parts. This coincides with Clarke's treatment of the subject in his letters to Leibniz. For Clarke (a) and its negation are both consistent with the law of contradiction. Moreover, the truth of (b) cannot be established without the additional appeal to the principle of sufficient reason or some interpretation of it. This means that if Clarke were to give 'satisfactory' answers (as he felt he ought to) to such questions as Why did composite substances come to exist in space rather than remain an ideal possibility? and Why are they composed of simple parts rather than of infinitely divisible parts? then he has to appeal to some version of the principle of sufficient reason. According to his version of the principle this reason is the mere Will of God. In reply to Leibniz's employment of the principle of sufficient reason to deny the actuality of indiscernible material atoms (simple parts) Clarke says:

In which case, the parts of space being alike, 'tis evident there can be no reason, but mere will, for not having originally transposed their situations. And yet even this cannot be reasonably said to be a will without motive; for as much as the wise reasons God may possibly have to create many particles of matter exactly alike, must consequently be a motive to him to take (what a balance could not do,) one out of two absolutely indifferents; that is, to place them in one situation, when the transposing of them could not but have been exactly alike good.[15]

Without some such appeal to a version of the principle of sufficient reason the thesis can only hold that if composite substances were actual then they would be made up of simple parts but it cannot proceed to hold, as it does, that composite substances are actually made up of simple parts. The thesis states: 'Nothing any where exists save the simple or what is composed of the simple.' This means, in addition, that for Clarke the indiscernibility of two 'things' does not prevent God from acting as Leibniz would hold. Clarke writes:

And even in compounds, there is no impossibility for God to make two drops of water exactly alike. And if he should make them exactly alike, yet they would never the more become one and the same drop of water, because they were alike. Nor would the place of the one, be the place of

---

[15] Fifth letter, paras. 1–20. See also paras. 21–5.

the other; though it was absolutely indifferent, which was placed in which place.[16]

To round up the Newtonian picture of the world embodied in the thesis and its proof Kant discusses the implications they have concerning the traditional idea of the soul as a simple mental substance. The statement of this implication occurs in the observation on the antithesis (A443–B471). Its refutation, from the Leibnizian point of view, occurs mainly in the proof of the antithesis. At this point I will not concern myself with the refutation. Kant says in the observation on the antithesis that the thesis implies the assertion 'that the object of inner sense, the "I" which there thinks, is an absolutely simple substance' (A443–B471).

The question of the soul as a simple mental substance was sharply debated by Clarke and Leibniz. The latter accused his English correspondent (and thus the Newtonians) of adopting a simplistic form of Cartesian dualism whereby 'reality' is divided into atomic material particles, on the one hand, and simple perceiving mental substances (souls), on the other. In addition, the Newtonians uncritically assume that the two realms of matter and spirit influence each other and are combined in man in some mysterious and inexplicable fashion.[17] Clarke answers the charge of materialism urged by Leibniz against the Newtonians in the following manner:

That some make the souls of men, and others even God himself to be a corporeal being; is also very true: but those

[16] Fourth letter, paras. 3 and 4.

[17] According to Clarke 'A soul is part of a compound, whereof body is the other part; and they mutually affect each other, as parts of the same whole.' (Second letter, para. 12.)

who do so, are the great enemies of the mathematical principles of philosophy; which principles, and which alone, prove matter, or body, to be the smallest and most inconsiderable part of the universe.[18]

Leibniz criticizes this conception of the soul and its relationship to the body by raising familiar difficulties:

Besides, the soul being indivisible, its immediate presence, which may be imagined in the body, would only be in one point. How then could it perceive what happens out of that point? I pretend to be the first, who has shown how the soul perceives what passes in the body.[19]

Again he says:

But the soul having no immediate influence over the body, nor the body over the soul; their mutual correspondence cannot be explained by their being present to each other.[20]

Clarke replies by restating the dualistic position accepted by the Newtonians:

The soul's being indivisible, does not prove it to be present only in a mere point. . . . The soul likewise, (within its narrow sphere,) not by its simple presence, but by its being a living substance, perceives the images to which it is present; and which, without being present to them, it could not perceive.[21]

Retaining in mind these considerations about mental substance in connection with the thesis will help us later on in understanding certain apparently obscure passages and arguments in the proof of the antithesis. Furthermore, it should be clear now that the common

[18] First letter, para. 1.
[19] Second letter, para. 4.
[20] Ibid., para. 5.
[21] Ibid., paras. 4 and 5.

assumption that the second antinomy deals with the constitution of material substance only is quite inaccurate.
In addition to the views of the Newtonians about
material substance found in the thesis (and its proof) the
antinomy considers also their complementary views
about the other kind of substance, viz. the simple
perceiving soul.

Before moving on to consider the antithesis and its
proof I shall connect the second antinomy with the
Dissertation of 1770 as a whole. According to 'the
principles of the form of the sensible world' (i.e.
three-dimensional space) as expounded in the *Dissertation* the idea of a composite material substance infinitely
divisible to smaller and smaller parts is impossible by
the very definition of the nature of compositeness in
such a space. And this is precisely the contention of the
thesis and its proof about the nature of compositeness.
On the other hand, according to 'the principles of the
form of the intelligible world' the whole idea of the
thesis and its proof is based on 'a false principle,
borrowed from sensitive knowledge, namely, that in such
a compound there cannot be an infinite regress in the
composition of parts . . .' (79). This makes the idea of the
infinite divisibility of matter not only legitimate but, in
effect, the only really 'acceptable' view. And this is
precisely the claim of the antithesis.

III

The antithesis of the second antinomy states that:

No composite thing in the world is made up of simple
parts, and there nowhere exists in the world anything
simple.

The proof of the antithesis is really made up of two parts. I shall refer to the first part as P_1 (A435–B463) and to the second part as P_2 (A437–B465). Following is P. F. Strawson's neat summary of P_1:

On behalf of the antithesis he argues, in effect, that what is composite and space-occupying can be made up only of parts which are themselves space-occupying and that everything space-occupying is extended and therefore composite; hence that the composite cannot be made up of non-composite or simple parts.[22]

The basic procedure followed in P_1 is to assume the claim of the thesis about the constitution of space-occupying material objects and then to raise a crucial question about the nature of the relationship holding between the parts constituting such objects and the parts of the space they occupy. P_1 tries to show that on the assumption of the thesis any account of this relationship will lead to absurdities and contradictions which have the effect of discrediting the initial assumption itself. From the assumption of the truth of the thesis 'it follows that the simple would be a composite of substances—which is self-contradictory'.

The general idea underlying this procedure may be summarized as follows: since the alleged simple parts making up a composite material object are themselves in space then they are so in exactly the same sense as the composite object itself is said to be in space. This means that the alleged simple parts are just as composite as the ordinary object occupying space. From the point of view of the antithesis to be in space is to be composite and no privileged cases (such as Newtonian simple parts in space) are allowed.

[22] *The Bounds of Sense*, p. 183.

It should be immediately noted that from the point of view of the antithesis (Leibnizian metaphysics) no separation is possible between the nature of the space-occupying material substances and the spatial container itself. According to this point of view the finite or infinite divisibility of space-occupying matter necessarily entails the finite or infinite divisibility of container-space itself. On this theory the idea of the divisibility of space beyond the matter occupying it is as absurd as the idea of the divisibility of matter itself beyond the divisibility of its spatial-container. This organic interdependence between space and matter is the Leibnizian idea underlying the refutation of the thesis in P_1. On this basis, only, it is possible for P_1 to argue from 'the continuous nature of space to the continuous nature of the matter which occupies it'.[23] This idea was urged by Leibniz against Clarke who, as a good Newtonian, held that space itself was absolutely indivisible, while material substances in space are finitely divisible on their own.

Clarke wrote:

Nor is there any difficulty in what is here alleged about space having parts. For infinite space is one, absolutely and essentially indivisible: and to suppose it parted, is a contradiction in terms. . . .[24]

Leibniz disagrees about the indivisibility of the spatial container:

But 'tis sufficient that space has parts, whether those parts be separable or not; and they may be assigned in space, either by the bodies that are in it, or by lines and surfaces that may be drawn and described in it.[25]

[23] Norman Kemp Smith, *Commentary*, p. 490.

[24] Third letter, para. 3. See also his second letter, para. 4, and fourth letter, paras. 11 and 12.

[25] Fifth letter, para. 51.

It becomes clear, then, that the intention behind the argument in P_1 is twofold. (*a*) The attempt to show that the idea of an indivisible space containing divisible material things is absurd. The reason for this absurdity may be summarized in the following fashion: if an object (*o*) is contained in a space (*s*), then the parts constituting (*o*) would have to be contained (located) in the parts of (*s*). But on the Newtonian assumption of the absolute indivisibility of space this would be impossible, since absolute space has no 'real' parts which can be said to contain the 'real' parts of the composite objects in space. If the defenders of the thesis wish to insist on the absolute indivisibility of space then there can be no 'simplest part' of matter since by definition there can be no 'smallest part' of space to contain it. Consequently, we would have to think that the simplest part of matter is located nowhere which is absurd from the point of view of the thesis.

(*b*) The assertion of the infinite divisibility of space and matter after the Leibnizian fashion. In other words, once the divisibility of material substances has been admitted no sufficient reason can be produced, according to Leibniz, to stop the process of subdivision at any one point rather than any other.

Accordingly the argument in P_1 may be construed as saying that 'since everything real, which occupies a space, contains in itself a manifold of constituents external to one another, and is therefore composite'; it would be simply arbitrary (and thus contrary to the principle of sufficient reason) to produce suddenly a privileged case of a 'real thing in space' which 'does not contain in itself a manifold of constituents external to one another', viz. a Newtonian simple part. Either the alleged simple part is really a composite 'made up of substances' or a violation of the principle of sufficient reason would have occurred. Since the latter cannot occur then there are no simple parts.

Now, it is well known that Leibniz rejected the reality of simple parts (atomic particles) in the correspondence with Clarke on the grounds that their existence would constitute a violation of the principle of sufficient reason as he understood it and of the principle of the identity of indiscernibles as he derived it from the principle of sufficient reason itself. Leibniz states his position thus:

And the case is the same with atoms: what reason can any one assign for confining nature in the progression of sub-division? These (i.e. all the items in the atomistic theory) are fictions merely arbitrary, and unworthy of true philo-sophy.[26]

His full position on the matter is identical with the claim of the antithesis. He says:

The least corpuscle is actually subdivided *in infinitum*, and contains a world of other creatures, which would be wanting in the universe, if that corpuscle was an atom, that is, a body of one entire piece without subdivision.[27]

It should be also noted that the antithesis views the relation of compositeness in exactly the same manner as Leibniz did, i.e. as neither external nor accidental. According to the point of view of the antithesis: 'A real composite is not made up of accidents (for accidents could not exist outside one another, in the absence of substance) but of substances. . . .'

The second and longer part of the proof of the anti-thesis[28] (P_2) is also directed against the claim that there

[26] [27] PS. to his fourth letter.
[28] A437–B465.

are such things as simple substances. However, we should note that some new concepts, which we did not quite encounter in the earlier sections of the antinomy, creep into this part of the proof. For example, such expressions as 'inner and outer perception' and 'possible experience' appear in P_2.

P_2 begins with a comment on that segment of the anti-thesis which states that 'nowhere in the world does there exist anything simple'. Then it informs us that:

. . . the existence of the absolutely simple cannot be established by any experience or perception, *either outer or inner*;[29] and that the absolutely simple is therefore a mere idea, the objective reality of which can never be shown in any possible experience, and which, as being without any object, has no application in the explanation of the appearances.

I think the argument of P_2 aims at performing two tasks. (*a*) It will be remembered that the proof of the thesis ended with the remark that the actual isolation of the simple parts constituting composite material sub-stances is impossible. P_2 admits this truth by saying that the existence of the absolutely simple cannot be estab-lished by any outer experience or perception. Then it utilizes this admission to discredit the claims of the thesis on the grounds that whatever cannot be, in prin-ciple, the object of such experience or perception is nothing but 'a mere idea' which corresponds to no objective reality in the world. Such a mere idea 'being without an object, has no application in the explanation of the appearances' (A437–B465).

This specific refutation of the claim of the thesis is essentially an employment of a version of Leibniz's

[29] My italics.

principle of the identity of indiscernibles as it appears in the correspondence. Leibniz maintained that if, in principle, there is no way of distinguishing between two states of affairs that are alleged to be distinct then there is in fact one and the same state of affairs. He employed this principle to show that Clarke's contention that the whole universe might be moved in space without a discernible difference appearing in it is without any meaning.[30] P_2 is, in effect, arguing that when the Newtonians admit that the simple parts of matter, although real and ultimate, are completely 'insensible' (in Clarke's words) or can never be the objects of 'outer perception or experience' (in Kantian terms) then they are depriving themselves of the only possible way which can make the state of affairs they are describing 'really' and meaningfully distinct from the state of affairs described by their opponents.

In P_2 we find, in effect, a restatement of the Leibnizian position on the question of the reality of 'insensible' atomic particles. Leibniz wrote in reply to Clarke:

But the author opposes this consequence, because (says he) sensible bodies are compounded; whereas he maintains there are insensible bodies, which are simple. I answer again, that I don't admit simple bodies. There is nothing simple, in my opinion, but true monads, which have neither parts nor extension.[31]

It is also interesting to note that the description of the doctrine of simple substances in P_2 as 'a mere idea' is strongly reminiscent of Leibniz's constant references, in the correspondence, to Newtonian atomism and its

[30] *The Leibniz–Clarke Correspondence*, p. xxiii.
[31] Fifth letter, para. 24.

implications as 'a fiction', 'a fancy', and 'a pleasing imagination'.

(b) The reference we find in P₂ to inner perception or experience is directed against the additional Newtonian claim that simple mental substances (the thinking I) exist. Such a substance is supposedly given (or known) in inner perception or experience in its naked simplicity. It is well known that traditional arguments in favour of the reality of a simple mental substance have always had to rest on some special inner experience or perception. In contrast, the arguments supporting the reality of simple material substances did not find themselves constrained to appeal to an 'outer experience' of these simples. On another occasion Kant summarized the position of the Newtonians (who insist on the reality of absolute time) concerning mental substance in the following words:

On the other hand, the reality of the object of our inner sense (the reality of myself and my state) is, [they argue,] immediately evident through consciousness. The former (i.e., external objects) may be merely an illusion; the latter is, on their view, undeniably something real. (B55)

Now, P₂ argues that this sort of experience of a simple mental substance is impossible. This would eliminate the remaining piece of alleged evidence in favour of the thesis. This aspect of P₂ is directly related to the last segment of the observation on the antithesis (A443–B471) where the question of the reality of a simple mental substance is openly discussed in the antinomy. In this segment of the observation Kant says that the question 'has been fully considered above' and the 'above', here, has no passage to refer to in the antinomy except P₂.

In other words, P_1 and P_2 taken together would have the effect of completely banishing Newtonian simple substances (of whatever sort they may be) from the realm of the actual and the real. This is expressed in P_2 as follows:

Whereas the first proposition (i.e., the first part of the statement of the antithesis and its proof) banishes the simple only from the intuition of the composite, the second excludes it from the whole of nature. Accordingly it has not been possible to prove this second proposition by reference to the concept of a given object of outer intuition (of the composite), but only by reference to its relation to possible experience in general.

The rejection, in the antithesis, of the reality of simple mental substances is really an expression of Leibniz's stand on the issue in opposition to Clarke. Leibniz, as pointed out above, accused the Newtonians (in Clarke's person) of adopting 'the vulgar notion' concerning the nature of the soul and its relationship to the body.[32] He ridicules this vulgar and simplistic conception in such terms as these:

To say that it (the soul) is diffused all over the body, is to make it extended and divisible. To say it is, the whole of it, in every part of the body, is to make it divided from itself. To fix it to a point, to diffuse it all over many points, are only abusive expressions, *idola tribus*.[33]

Now, I would like to deal with the main argument and contents of P_2. Kant states in the observation on the antithesis that P_2 is meant to stand against a certain

[32] Third letter, para. 12.
[33] Ibid. In fact Leibniz distinguishes the modern Christian materialists (who believe in the mathematical principles of natural philosophy) from the ancient ones by saying that the former admit not only the reality of bodies but also the reality of immaterial substances (second letter, para. 1).

dogmatic assertion implied by the thesis which has the characteristic of being:

. . . the only one of all the pseudo-rational assertions that undertakes to afford manifest evidence, in an empirical object, of the reality of that which we have been ascribing only to transcendental idea, namely, the absolute simplicity of substance—I refer to the assertion that the object of inner sense, the 'I' which there thinks, is an absolutely simple substance. (A443–B471)

The refutation in P₂ of the claim that the manifest evidence of inner sense proves the reality of the simple mental substance may be summarized in the following points:

1. Assume that there is given in experience an object which is absolutely simple.

2. Such an object would have to be 'known (experienced, empirically intuited) as one that contains no manifold [factors] external to one another and combined into unity'.

3. However, the fact that no perception of such manifoldness in the object occurs does not necessarily mean that the object is absolutely free from all kind of manifoldness whatsoever.

4. The absolute simplicity of such an object can be conclusively established if and only if the absolute absence of such manifoldness in the object is indubitally demonstrated.

5. 'It follows that such simplicity cannot be inferred from any perception whatsoever' (inner or outer) because no such indubitable demonstration is ever possible on empirical or experiential grounds.

6. Therefore, '. . . since by the world of sense we

must mean the sum of all possible experiences, it follows that nothing simple is to be found any where in it'.

This sort of argument serves to affirm again Kant's general claim that no appeal to experience can be of any real service in satisfactorily settling the dispute between the claims and counter-claims of the antinomy. The attempt of the partisans of the thesis to demonstrate, at least, the reality of one type of simple substance by appealing to an immediate inner experience is quickly neutralized by the counter claim of the antithesis that in the very nature of the case no experience of an object as simple (non-composite) can definitively lead us to the conclusion that the object *itself* is actually simple and non-composite. The opposing claims of the antinomy pretend to be truths not about our modes of experiencing objects (ourselves) but about the nature of things as they are in themselves. That is they pretend to give metaphysical knowledge about the nature of reality as such. This knowledge supposedly possesses the demonstrative certainty and assurance befitting metaphysical information about some aspect of the world. Consequently, no appeal to experience on behalf of either the thesis or antithesis is permissible in this case since experience can yield only contingent results devoid of the sort of certainty and indubitability which metaphysical claims are supposed to possess.

We may note here that the heart of the conflict between the two sides of the antinomy may be regarded in terms of how each side interprets the notion of possible experience. The partisans of the thesis agree in general that experience cannot yield definitive knowledge of whether an object is simple or manifold. But then they proceed to lay down a restriction on this general affirma-

tion by saying that there is at least one privileged case
of an experience which genuinely yields knowledge of
an absolutely and really simple object. This privileged
experience is the inner perception of the self as a simple
substance. Now, from the point of view of the antithesis
the erection of such a privileged experience would seem
arbitrary. It is called for by no convincing 'sufficient
reason'. It can be readily seen also that this specific conflict in the second antinomy is a reflection of the dispute
between Leibniz and Clarke over the interpretation of
the principle of sufficient reason. Both correspondents
agreed to the truth of the principle in general, but
Clarke sought, in effect, to set up a privileged instance
to which the principle did not quite apply in the way
that he applied it to all other things. This privileged
case was, of course, the Will of God. Leibniz, naturally,
objected that such a procedure on the part of Clarke
was arbitrary and irrational. For he held that since the
principle was true, then it must be applied unrestrictedly
and with no exception whatsoever.

Again, the notion of 'possible experience' as it occurs
in P_2 is identical with the idea of 'actuality' as distinguished from pure possibility determined by the law of
contradiction. Both Clarke and Leibniz maintained
that the contents of actuality are determined according
to the principle of sufficient reason. However, Clarke
held that these contents include simple substances. He
also held that at least inner sense indubitably reveals
the reality of one such substance. On the other hand,
Leibniz held that the principle of sufficient reason determined the contents of actuality in such a way that no
simples are actual. They remain ideal possibilities and
hence no experience, inner or outer, can have for its

contents the simple as such. The argument in P_2 is, then, basically a retraversing of the grounds which led Leibniz to the rejection of the reality of simple substances for being inconsistent with the principle of sufficient reason. He argued that the actuality of simple substances would not be agreeable to God's wisdom.[34]

In the observation on the antithesis Kant produces an explanation as to how the illusion of a single substance given in inner perception arises. This illusory notion, according to Kant, is the product of a certain mistaken, but traditional, train of thought. We can summarize it as follows: it is always possible for us to form the general idea of an 'object' by abstracting from each and every one of its concrete determinations as they are given in 'empirical intuition'. Having subjected the object to this abstractive operation from every 'synthetic determination of its intuition' we naturally remain with the representation of the object as a simple entity. This is what happens in the case of the simple 'I' which is supposedly given in inner perception. In thinking of our states we abstract from all the concrete determinations which inner sense involves in their perception. This abstractive operation leads us to think that what remains is the bare 'I' void of all manifoldness (compositeness) since all its concrete determinations have been removed or thought away. Then the additional claim is made that a special inner experience does actually reveal this non-composite entity in its bare simplicity.

Kant observes that as long as we think about the self

[34] Leibniz's fifth letter, para. 30. To be noted here is that the famous monads of Leibniz with their principle of activity are quite a different sort of thing from the inert simples of the Newtonians against which Leibniz argued so vehemently in the correspondence.

in terms of this misleading abstractive train of thought we will form to ourselves 'the quite bare representation "I" in isolation from the concrete determinations which inner sense necessarily involves. Once we avoid this misleading habit of thought it would become clear to us that: 'as an object of intuition, it (the "I") must exhibit [some sort of] compositeness in its appearance; and it must always be viewed in this way if we wish to know whether or not there be in it a manifold [of elements] *external* to one another'.[35]

I shall end my discussion of the antithesis by commenting on criticisms urged by Norman Kemp Smith against Kant's handling of the two sides of the antinomy. Kemp Smith accuses Kant of having stated the two sides of the alleged antinomy in such a way as to have emptied it of the antinomial conflict supposedly obtaining between the thesis and the antithesis. He writes:

But as the thesis and the antithesis thus refer to different realities, the former to things in themselves conceived by pure understanding, and the latter to the sensuous, no antinomy has been shown to subsist.[36]

I have tried to show that the thesis and antithesis do not deal with different realities at all but with the nature and constitution of substance material and mental. Furthermore, both thesis and antithesis, from the critical point of view, make dogmatic metaphysical claims about the nature of things as they are in themselves. Consequently they lead to a conflict which can be resolved neither empirically nor rationally. Both sides of the antinomy claim to be informing us, with demonstrative certainty, what the nature of things 'really' is,

---

[35] Observation on the antithesis.

[36] *Commentary*, pp. 490–1.

and in which way the world is 'really' constituted. The Newtonians were as dogmatic and confident about the truth of their version of how things really are as the Leibnizian and Wolffian metaphysicians. Kant has indeed succeeded in constructing an antinomy by presenting us with these two opposed systems of ideas and arguments about the question of the ultimate simplicity versus the ultimate compositeness of substances in the world.

I shall mention another mistaken remark registered by Kemp Smith. Speaking of the antithesis and its proof he writes:

> The Leibnizian standpoint is here completely deserted. Instead of proceeding to demonstrate the direct opposite of the thesis, Kant in this argument deals with the extended bodies of empirical intuition.[37]

I hope to have convincingly argued that the Leibnizian standpoint is adhered to in the antithesis and its proof. Furthermore, Kemp Smith's comment about the 'extended bodies of empirical intuition' is irrelevant since the very conflict between thesis and antithesis is centred on whether the extended bodies in space are ultimately simple or ultimately composite.

IV

Those parts of the observation on the antithesis which do not deal with mental substance deserve some separate discussion. The observation opens with a remark about the sort of objections urged by atomists[38] (Newtonians) against the Leibnizian doctrine of the infinite divisibility

[37] *Commentary*, p. 490.
[38] Called monadists in this context.

of matter. Kant says that the proof of this doctrine is purely mathematical. However, Kant continues, the sort of objections urged by the atomists against the doctrine of infinite divisibility and its mathematical proof subjects the atomists themselves to certain suspicions. These have to do with certain aspects of the standpoint of the atomists and with the soundness of their conception of the nature of mathematical proof. The trouble with the atomists, in this connection, is formulated by Kant in this way:

> For however evident mathematical proofs may be, they (the atomists) decline to recognize that the proofs are based upon insight into the constitution of space in so far as space is in actual fact the formal condition of the possibility of all matter. They regard them merely as inferences from abstract but arbitrary concepts, and so as not being applicable to real things.[39]

In explaining the significance of Kant's diagnosis it would be convenient to limit our discussion to the traditional science of geometry for Kant's primary concern is obviously space and the things in space.

Kant is here reaffirming one of his basic criticisms of the Newtonian conception of the nature of geometry. This conception remains unable to guarantee the objective validity and universal applicability of the propositions of geometry to appearances (things) in space. The Newtonians adopted an axiomatic view of geometry in opposition to the Leibnizian conception which sought to derive all the propositions of this science from the law of contradiction plus certain definitions. According to the latter view there are no genuine indemonstrable axioms in geometry and there can be only one system

[39] Observation on the antithesis (A439–B467).

of geometry which is free from contradiction. This is the Euclidean geometry which describes with *a priori* certainty the properties of space and of objects in it.[40]

Now the Newtonians rejected this theory of geometry in favour of an axiomatic view which holds that the theorems of geometry cannot be demonstrated without employing certain genuine indemonstrable axioms in addition to the law of contradiction and the definition of the basic concepts of the science. It follows that if certain appropriate changes are introduced into the axioms other kinds of geometries will result which are free from contradiction but which have no application to things (no objective validity). In other words, Kant is saying that on account of this axiomatic point of view the Newtonians fall into the trap (whether they recognize it or not) of regarding the truths of geometry 'as inferences from abstract but arbitrary concepts and so as not being applicable to real things'. Kant had already made the same point in the *Dissertation* where he refers to the arbitrariness of mathematics which, according to this view, would be 'put together only for the purpose of deducing consequences from it' (38).

Kant's basic complaint in the observation on the antithesis is that on the axiomatic point of view adopted

---

[40] According to Leibniz 'The great foundation of mathematics is the principle of contradiction, or identity. . . . This single principle is sufficient to demonstrate every part of arithmetic and geometry, that is, all mathematical principles.' (Second letter to Clarke, para. 1.)

The consequence of this view is that there are no genuine indemonstrable axioms in geometry. Leibniz writes: 'This is more or less the case with those propositions which are generally regarded as axioms, but which, like the theorems could be proved and are worth proving; nevertheless they are permitted to function as axioms as though they were original truths.' (G. Martin, *Leibniz, Logic and Metaphysics*, Manchester University Press, 1964, p. 70.)

by the atomists the admitted universal and necessary applicability of geometry to space and the objects in it is reduced to an inexplicable mystery. At best, the objective validity of Euclidean geometry would be purely accidental. Kant is naturally very anxious to reform the axiomatic view of geometry to prevent it from leading to such embarrassing results. If 'philosophy is not to play tricks with mathematics'[41] it has to show that the truths of Euclidean geometry are not mere inferences from abstract and arbitrary concepts, which may or may not be applicable to 'real things' in space (appearances in space). The truths of geometry must be made to rest upon firmer foundations which guarantee their universal applicability and objective validity for all things in space. This is exactly the sort of reform which Kant's critical theory of space and geometry is intended to produce. Without such a reform the axiomatic point of view, 'uncritically' adopted by the atomists, quickly degenerates into an attempt 'to reason away by sophistical manipulation of purely discursive concepts the evident demonstrated truth of mathematics'.[42]

The critical philosophy achieves this piece of reform by holding that the axioms of Euclidean geometry are not abstract and arbitrary concepts put together for the sake of making inferences from them (as is the case with non-Euclidean geometries) but that they are based on an 'insight into the constitution of space' as the form of human sensibility (or of all outer intuitions).[43] Since this form of sensibility is also the form of all 'outer' appearances the objective validity and universal applicability

[41] [42] Observation on the antithesis.

[43] Kant explains the nature of this reform in the *Dissertation* of 1770. See sect. ii, para. 12 and sect. iii, para. 15c.

of the truths of geometry to all appearances in space is fully guaranteed. Thus, Kant distinguishes in the *Dissertation* between 'intuitively given hypotheses' (axioms) and 'invented hypotheses' (63).

Now we return to the text of the observation on the antithesis. Having taken to task the Newtonians for not seeing the need for reforming their axiomatic point of view about mathematics along critical lines Kant presses his criticism by throwing two rhetorical questions at the atomists. The significance of the two questions is that on the unreformed axiomatic point of view it should be possible 'to invent a different kind of intuition[44] from that given in the original intuition of space' which is a ridiculous consequence. Again, on the unreformed axiomatic point of view 'the *a priori* determinations of space (could) fail to be directly applicable to what is only possible in so far as it fills this space'. This is also a ridiculous result which works to show the untenability of the view accepted by the atomists. Naturally the implication is that none of these ridiculous results would follow if we were to adopt the critical theory of space and geometry. Kant, then, proceeds to point out, in the observation, a certain 'absurdity' entailed by the views of the atomists about absolute space and its relationship to the matter located in it. This absurdity is further grounds for not accepting the Newtonian picture of things without the qualifications and reforms added to it by the critical theory. According to Kant, were we to listen to the atomists then beside the simple mathematical points (which are not the parts but the limits of a space)

[44] Such as an intuition which corresponds to non-Euclidean axioms.

we would have to accept the strange conception of simple dimensionless physical points which are parts of space and which come to fill a space through their mere aggregation.

Kant, then, proceeds to reaffirm a familiar point in the observation (A441–B469). The philosophers (basically the Newtonians) who play such tricks with mathematics do so because they do not realize that geometry deals only with *appearances* and their condition (i.e. the spatial form of sensibility).[45] Only on such a critical view can we put an end to these tricks after exposing them. As a result geometry would then rest on firm foundations as an *a priori* science which is at the same time universally applicable and objectively valid *vis-à-vis* all outer phenomena.

Kant continues his explanations in the observation by saying that the partisans of the thesis and antithesis among philosophers claim to be talking not about appearances, but about things as they are in themselves. Consequently it is not enough for the former merely to assert the abstract concept of the simple in opposition to the concept of the composite in order to show the truth of their doctrine. If the Newtonians are to settle the dispute in their favour 'what has to be found is an intuition of the simple for the intuition of the composite (matter)'. In other words, to oppose the concept of the simple to the 'pure concept of the composite formed by

[45] This could easily apply to the Leibnizians as well. Their theory of mathematics was severely criticized by Kant in the *Dissertation* and the *Aesthetic*. For example, Kant thought that the relational theory of space and time obliges the Leibnizians 'to deny that *a priori* mathematical doctrines have any validity in respect of real things (for instance, in space), or at least to deny their apodeictic certainty' (A40–B57).

the understanding' serves only to lead to a hopeless dialectical conflict. However, concerning the possibility of obtaining such an intuition of the simple Kant says: 'But by the laws of sensibility, and therefore in objects of the senses, this is quite impossible.' It is impossible because all sensuous intuition occurs through the two forms of sensibility which by their very nature entail manifoldness and diversity. Concerning the objects of sense the perception of the really simple is impossible. Kant's general position on this matter is clearly stated in the first paragraph of the first section of the *Dissertation* of 1770. He says for instance:

For it is one thing, given the parts, to conceive the *composition* of the whole by an abstract notion of the intellect, but another thing to take this general notion as a problem set by the reason, and to *follow it out* by the sensitive faculty of apprehension, i.e., to represent it in the concrete by distinct intuition.

Thinking abstractly about the nature of composite substances we come to the abstract notion of simples prior to all composition. But in so far as we are dealing with appearances in space it would be impossible 'to follow out the simple by the sensitive faculty of apprehension and to represent it in the concrete by distinct intuition'. The reason for this is that 'empirical intuition in space carries with it the necessary characteristic that no part of it is simple because no part of space is simple'.[46] Then, Kant makes the following comment in the observation:

The monadists have, indeed, been sufficiently acute to seek escape from this difficulty by refusing to treat space as

[46] Observation on the antithesis.

a condition of the possibility of the objects of outer intuition (bodies), and by taking instead these and the dynamical relation of substances as the condition of the possibility of space.

Kant is saying that the atomists are aware of the difficulties which their theory of space involves. These difficulties have to do with their conception of the nature of the science of geometry and of the relationship holding between space and the matter in it. However, they reject the 'critical' solution of the problem, i.e. they refuse 'to treat space as a condition of the possibility of the objects of outer intuition (bodies)'. This could apply equally well to the monadists of the Leibnizian (in the strict sense) as well as of the Newtonian variety. For the Leibnizians space is dependent upon the existence of 'the objects of outer intuition' as is commonly known. On the other hand, the Newtonians argue that certain dynamical phenomena require a distinction between absolute and relative acceleration, i.e. these phenomena require a distinction between absolute and relative motion which presupposes the reality of absolute space and time. In other words, these monadists argue from the dynamical relations of matter to the necessity of absolute space.

Examples of atomists who argued primarily from the existence of bodies and their dynamical qualities to the reality of absolute space are Clarke and Euler. Their argument moves from the study of motion and the need to distinguish absolute from relative motion to the postulation of the reality of absolute space. Kant once wrote about Euler: '[He] brings to view the difficulty of assigning to the most general laws of motion a determinate meaning, should we assume no other concept of space than that obtained by abstraction from the relation of actual things' (21).

Euler, as an interpreter and defender of the Newtonian system of nature, argued that since the law of inertia holds that a body not acted on by any external forces will cover absolutely equal distances in absolutely equal times, it will have to be the only body in the universe describing such a motion. For the presence of other bodies would act on it in accordance with the universal law of gravitation. The law of inertia presupposes, then, a meaning for the idea of the motion of a single body covering equal intervals of space in equal intervals of time. No theory of space could endow such an idea with meaning except the theory of absolute space. The conclusion of this argument is that the validity of Newton's mechanics and the meaningfulness of their concepts presuppose the reality of absolute space as the container of bodies and their motions.

Clarke also presented similar arguments in combating the opinions of Leibniz. For instance he says:

If the material universe CAN possibly, by the will of God, be finite and moveable; . . . then space, (in which that motion is Performed,) is manifestly independent upon matter.[47]

Again, in arguing against Leibniz's conception of the motion of a body as always relative to other bodies, Clarke brings out Euler's point about the motion of a single body in space. He indicates that on Leibniz's view the motion of such a body would be impossible and this is absurd (for a Newtonian): '. . . and yet no way is shown to avoid this absurd consequence, that then the mobility of one body depends on the existence of other

[47] Fifth letter, paras. 73-5.

bodies; and that any single body existing alone, would be incapable of motion . . .'.[48]

Kant considers this manner of avoiding the difficulties involved in the unreformed Newtonian point of view to be no more than an evasion. He points out in the observation that he has already refuted this evasive attempt in the *Aesthetic*. This part of the observation ends with the following statement: 'This evasion of the issue is therefore futile, and has already been sufficiently disposed of in the Transcendental Aesthetic. The argument of the monadists would indeed be valid if bodies were things in themselves.' Kant remains confident that the only successful way of overcoming these difficulties is by adopting the critical theory according to which 'we have a concept of bodies only as appearances; and as such they necessarily presuppose space as the condition of the possibility of all outer appearance'.[49]

I should like to end by pointing to some misleading remarks made by Norman Kemp Smith on the observation of the antithesis. Kemp Smith writes:

Kant, as already noted, argues in the *Observation* to this antithesis that all attempts 'made by the monadists' to refute the mathematical proof of the infinite divisibility of matter are quite futile, and are due to their forgetting that in this discussion we are concerned only with appearances.[50]

Kant is not arguing at all that the refutations of the mathematical proof, on the part of the Newtonians, are futile. He is saying that the sort of objections that the atomists bring against this mathematical proof reveal certain basic inadequacies in their conception of the

[48] Ibid., paras. 26–32. See also his fourth reply, paras. 13 and 14.
[49] Observation on the antithesis.
[50] *Commentary*, p. 491.

nature of mathematics and its objective validity *vis-à-vis* the world. The inadequacies emanate from their theory of the reality of absolute space. The observation contains Kant's summary discussions of these inadequacies in the light of his own critical theory of scientific knowledge. That is to say the observation simply repeats a number of familiar Kantian views about the Newtonians. These views have already occurred in the *Dissertation* and the *Aesthetic*.

Kemp Smith finds the remark made by Kant in the observation[51] about the Newtonians who argue from the dynamical relations of substances to the reality of absolute space inconsistent with the proof of the thesis![52] There is no inconsistency whatsoever between Kant's words in the observation and the proof of the thesis. The latter deals with the constitution of material substance in space and not with the constitution of space itself. Kant's words in the observation point to the failure of Newtonians to overcome the deficiencies of their theory along 'critical' lines and to the evasive attempts of some of them to remedy the deficiencies by arguing from the dynamical properties of matter to the necessary reality of absolute space.

Kemp Smith holds that in the observation Kant 'presents the mathematical proof of the continuity of matter as conclusive'.[53] There is no evidence that Kant entertained such a view. He was as dissatisfied with the Leibnizian conception of mathematical proof as he was with that of the Newtonians. In the *Aesthetic* he criticized both schools. Now the mathematical proof of the infinite divisibility of matter may be quite valid formally.

[51] The passage is quoted on pp. 80–1.
[52] [53] *Commentary*, p. 491.

But whether it proves anything about the nature of physical reality is a different matter. There is no evidence in the observation that Kant wanted to say that this proof was conclusive in the latter sense.

THE THIRD ANTINOMY

I

THE third antinomy deals with the problem of causality. The thesis holds that in the world there are two types of causality: (1) 'Causality in accordance with laws of nature' whereby events follow 'necessarily' from their antecedents according to a rule; (2) the causality of freedom which consists of the power of a cause to initiate an event absolutely independently from the antecedent state of the universe. This power is often described in the antinomy as 'an *absolute spontaneity* of the cause'.[1] It is also called transcendental freedom. The antithesis holds that there is only one kind of causality in the world and that is causality in accordance with the laws of nature. The so-called causality of freedom is an illusion.

Before I proceed to a detailed examination of the arguments and comments constituting the body of the antinomy I would like to note a certain interpretative remark made *à propos* of the subject of the third antinomy. Speaking of the two dynamical antinomies (i.e. the third and the fourth) T. W. Weldon states:

The conflict in fact is between the claim of the scientists that all natural events must be explicable by natural causes and that of the moralist that unless human action is spontaneous, and therefore undetermined by natural causes, the conceptions of obligation and desert become entirely without meaning.[2]

Weldon is, in effect, doing two things in this statement: (*a*) He is identifying the antithesis with the claims of the

[1] Proof of the thesis. Kant's italics.
[2] *Kant's Critique of Pure Reason*, p. 209.

scientists of the eighteenth century (who were mechanists in the main) and the thesis with the claims of moralists who have to worry about problems of obligation and deserts. (*b*) He is reducing the subject of the third antinomy to an essentially moral problem which has to do specifically with human actions and choices. I would like to argue that Weldon is inaccurate in his understanding of the topics treated in this antinomy.

The general sense of the thesis is much closer to the particular conception of the world-machine expounded by Newton and Clarke than it is to the opinions of any moral philosophers favouring free will. This conception holds that the world-machine not only has a First Cause but also requires His constant intervention into its workings in order to keep it running properly and on due course. That is to say, in addition to the natural mechanical causality governing the operations of the world-machine there is also the free causality of the Divine Sovereign and of human souls.

The claim of the antithesis is identical with Leibniz's doctrine of universal determinism. Leibniz states his point of view on this matter in the correspondence with Clarke. He says:

For the nature of things requires, that every event should have beforehand its proper conditions, requisites, and dispositions, the existence whereof makes the sufficient reason of such an event.[3]

Again, Leibniz restates the same deterministic position (couched in the language of 'freedom' as he interprets it). For example he writes:

For God, being moved by his supreme reason to choose, among many series of things or worlds possible, that, in

[3] Fifth letter, para. 18.

which free creatures should take such or such resolutions, though not without his concourse; has thereby rendered every event certain and determined once for all; without derogating thereby from the liberty of those creatures: that simple decree of choice, not at all changing, but only actualizing their free natures, which he saw in his ideas.[4]

Clarke replies to this notion of universal determinism by pointing out that this point of view leads to such 'terrible' things as fate, blind necessity, and materialism. Clarke thinks he can avoid these 'inconvenient' results by insisting on the need for an original free cause of the material universe.[5] He expounded his views, in replying to Leibniz, as follows:

The notion of the world's being a great machine, going on without the interposition of God, as a clock continues to go without the assistance of a clockmaker; is the notion of materialism and fate, and tends, (under pretence of making God a *supra-mundane intelligence*,) to exclude providence and God's government in reality out of the world. And by the same reason that a philosopher can represent all things going on from the beginning of the creation, without any government or interposition of providence; a sceptic will easily argue still farther backwards, and suppose that things have from eternity gone on (as they now do) without any

[4] Fifth letter, para. 6.

[5] Newton wrote on this subject the following: '. . . for we adore him as his servants; and a God without dominion, providence, and final causes, is nothing else but Fate and Nature. Blind metaphysical necessity, which is certainly the same always and everywhere, could produce no variety of things. All the diversity of natural things which we find suited to different times and places could arise from nothing but the ideas and will of a Being necessarily existing.' (From Newton's *Principia*, General Scholium, *The Leibniz–Clarke Correspondence*, p. 169.)

true creation or original author at all, but only what such arguers call all-wise and eternal nature.[6]

Another point to mention in this connection is Kant's view of the third antinomy as the dynamical counterpart of the first mathematical antinomy. The significance of this relationship is spelt out by Kant himself:

If you do not, as regards time, admit anything as being mathematically first in the world, there is no necessity, as regards causality, for seeking something that is dynamically first.[7]

In other words, those who take the side of the antithesis in the first mathematical antinomy will also take the side of the antithesis in the first dynamical antinomy and these are the Leibnizian metaphysicians as pointed out earlier on more than one occasion. Again, those who take the side of the thesis in the first mathematical antinomy (thus holding that the universe has a beginning in time) will also take the side of the thesis in the third antinomy concerning the dynamical beginning of the world and these are the Newtonians of that time.

Weldon's interpretation of the third antinomy as essentially treating a moral problem is too narrow. Kant's primary concern here is with a cosmological problem, i.e. the problem of causal determination and the two conflicting views prevalent about it. In more than one place in the *Critique of Pure Reason* he summarizes the essential conflict involved in the third antinomy in terms which leave no doubt about this fact. For example, he formulates the problem as having to do with 'whether there is generation and production

[6] First letter, para. 4.
[7] Observation on the antithesis. Kant's italics.

through freedom, or whether everything depends on the chain of events in the natural order . . . ' (A481–B509). Again, he restates the same conflict, in terms that show clearly his predominant scientific and cosmological interests:

When we are dealing with what happens there are only two kinds of causality conceivable by us; the causality is either according to *nature* or arises from *freedom*. (A532–B560)

A third formulation is as follows:

By freedom . . . in its cosmological meaning, I understand the power of beginning a state spontaneously. (A533–B561)

This does not mean, however, that the cosmological conflict about the nature of causality does not entail important consequences for moral questions such as the freedom of the will, obligation, rewards, and punishments, etc. In fact, this is clearly understood and clarified by Kant in his various comments on the antinomy. He carefully distinguishes transcendental from practical freedom and notes the nature of the relationship holding between them. The moral problem of practical freedom is only a special case of the more general problem of transcendental freedom. As a result 'the denial (with the antithesis) of transcendental freedom must, therefore, involve the elimination of all practical freedom' (A534–B562). Again Kant clarifies this point in the following words:

It should especially be noted that the practical concept of freedom is based on this *transcendental* idea, and that in the latter lies the real source of the difficulty by which the question of the possibility of freedom has always been beset. (A533–B561)

What is problematic about the question of practical freedom is precisely its transcendental and not moral aspect. If a satisfactory solution is found for the transcendental problem of freedom then the problem of the freedom of the will (human agents) may be considered, at least in principle, as also solved. Kant says in this connection:

What has always so greatly embarrassed speculative reason in dealing with the question of the freedom of the will, is its strictly transcendental aspect. The problem, properly viewed, is solely this: whether we must admit a power of *spontaneously* beginning a series of successive things or states.[8]

It is interesting to note also that freedom as a moral issue is brought up by Kant, in conjunction with the third antinomy, primarily when he is so deftly exploring the extra-rational reasons and motives which lead thinkers to rally around the thesis or the antithesis as the case may be. For example, among the reasons he mentions for the popularity of the thesis (as against the antithesis) are the moral, psychological, and religious reassurances provided to man by the seeming implications of the thesis that our will is free and its actions are voluntary.[9] After making an exposition of the practical and 'regulative' advantages and disadvantages of the thesis and antithesis, Kant reaches the conclusion that when we exclude all extra-rational reasons and motives in our consideration of the antinomial conflict then no conclusion can be rationally reached in favour of either side of the antinomy. If we look at the conflict for what it is worth and purely on its own merits then the final result cannot be more than hesitation and vacillation

[8] Observation on the thesis.
[9] This is brilliantly expounded and explored by Kant in A462–B490/A476–B504.

between the two opposed dogmatic claims. Kant says on this matter:

If men could free themselves from all such interests, and consider the assertions of reason irrespective of their consequences, solely in view of the intrinsic force of their grounds, and were the only way of escape from their perplexities to give adhesion to one or other of the opposing parties, their state would be one of continuous vacillation. (A475–B503)

II

The argument in favour of the thesis may be summarized in the following way:

1. Assume that 'there is no other causality than that in accordance with laws of nature'.

2. This means that 'everything which *takes place* presupposes a preceding state upon which it inevitably follows as a rule'.

3. Accordingly 'the causality of the cause' also presupposes 'a preceding state and its causality' and this applies to every member of the resulting series of causal states which *take place* in the world.

4. For a state (*c*) to be truly considered as the cause of another state (*e*) then (*c*) must contain in itself, *a priori*, the complete and sufficient explanation of (*e*). This is simply the meaning of the laws of nature as initially assumed.

5. Now, if the regressing series of causal states in the world were to go on indefinitely then, as the series will always remain incomplete, no truly complete and sufficient explanation would have been given of any member of that series (or of anything that *takes place* in the

world). This amounts to a violation of the meaning of the laws of nature and to a contradiction of our initial assumption.

6. It follows that the regressing series of causal states cannot remain incomplete. It ends with the causality of a state 'which is not itself determined, in accordance with necessary laws, by another cause antecedent to it'.

7. The causality of this state is 'an *absolute spontaneity*' whereby it really initiates 'a series of appearances, which proceed in accordance with laws of nature'.

It should be evident from this summary that the ruling idea of the argument is a certain interpretation of the meaning of the completeness or sufficiency of the explanatory conditions (or causal sequence) of a given event. Completeness here means that the enumeration of these explanatory conditions can be 'brought to conclusion in a finite and assignable time'. Otherwise, with such an enumeration 'no absolute totality of conditions determining causal relation can be obtained'[10] (as the antithesis holds). According to the thesis completeness means precisely the possibility of attaining such a synthetic totality, at least in principle. Accordingly any collection (or series) of such explanatory conditions must have an assignable beginning, end, and number. The proof of the thesis explains this meaning of completeness as follows:

If, therefore, everything takes place solely in accordance with laws of nature, there will always be only a relative and never a first beginning, and consequently no completeness of the series on the side of the causes that arise the one from the other.

[10] A533–B561.

This understanding of the completeness of a series of explanatory conditions parallels very nicely with the proof of the thesis of the first antinomy and with the Newtonian doctrine of the finiteness of the universe.

Granted that the series of causal explanations is finite and reaches its completion in a really first term the question arises as to the nature of the causality enjoyed by that term. Is it of the same nature as the mechanistic causality governing all events or is it of a different kind? The claim of the thesis is that the causality of the first term, unlike the mechanistic causality in nature, does not refer to any determining conditions outside itself. The causality of such a cause is 'not itself determined, in accordance with necessary laws, by another cause antecedent to it . . .'.[11] Without this sort of a first cause 'the series of appearances on the side of the cause can never be complete'.[12]

This view, embodied in the thesis and its proof, is really a restatement, in Kantian language, of the position expounded by Clarke in opposing Leibniz's thoroughly deterministic standpoint which allows no place for a privileged first cause. Clarke argued that if mechanistic causality were the only kind of causality in the universe then we will have only patients and no agent in the universe. For if an alleged agent does not contain in itself the power of initiating the action to be mechanically communicated to other things then it would not be a real agent (or initiating cause) but a mere patient. Clarke clarifies this point as follows:

. . . but the true and only question in philosophy concerning liberty, is, whether the immediate physical cause or principle of action be indeed in him whom we call the

[11] [12] Proof of the thesis.

agent; or whether it be some other reason sufficient, which is the real cause of the action, by operating upon the agent, and making him to be, not indeed an agent, but a mere patient.[13]

Clarke is, in effect, asserting the presence of two types of causality in the world: the mechanistic causality appropriate to all patients in nature and the free initiating causality of agents. The opposite point of view, represented by Leibniz and the antithesis, presents us with a world of patients only, mechanically an indefinitely communicating action to each other. Clarke complains against the latter deterministic scheme of things in the following words:

But indeed, all mere mechanical communications of motion, are not properly action, but mere passiveness, both in the bodies that impel, and that are impelled. Action, is the beginning of a motion where there was none before, from a principle of life or activity: and if God or man, or any living or active power, ever influences any thing in the material world; and every thing be not mere absolute mechanism; there must be a continual increase and decrease of the whole quantity of motion in the universe. Which this learned gentleman frequently denies.[14]

The claims of the thesis and its proof are no different from what Clarke is saying. According to the thesis this sort of mechanistic causality 'cannot, therefore, be regarded as the sole kind of causality' in the universe. Thus in Clarke's view there are two kinds of agents capable of initiating real action. The supreme first cause of the world-machine, for according to him 'things could not be at first produced by mechanism';[15] and the

[13] Fifth letter, paras. 1–20.
[14] Ibid., paras. 93–5.
[15] Ibid., paras. 110–16.

souls of human beings *in* the world. In other words, this kind of originating free causality is applicable to the origin of the world-machine and to certain actions and instances *in* the processes of the world-machine itself. Clarke states this point in the following way:

To suppose, that all the motions of our bodies are necessary, and caused entirely by mere mechanical impulses of matter, altogether independent on the soul; is what (I cannot but think) tends to introduce necessity and fate. It tends to make men be thought as mere machines, as Descartes imagined beasts to be; by taking away all arguments drawn from phenomena, that is, from the actions of men, to prove that there is any soul, or any thing more than mere matter in men at all.[16]

Kant reformulates Clarke's position in his statement of the case for the thesis of the antinomy in a manner which makes the argument less theologically loaded. For the thesis the first term of the series of causal conditions is an agent in Clarke's sense. This first agent supplies 'a first beginning, due to freedom, of a series of appearances' which 'make(s) an origin of the world conceivable'.[17] Kant then takes account of Clarke's second type of agents which are part of the world's processes. He begins by noting that although the thesis of the antinomy is primarily concerned with demonstrating the necessity of a first agent we may go beyond that 'to admit within the course of the world different series as capable in their causality of beginning of themselves, and so to attribute to their substances a power of acting from freedom'.[18] The example given by Kant to illustrate

[16] Fifth letter, para. 92.
[17] [18] Observation on the thesis.

his meaning is a simple spontaneous human choice. His example goes as follows:

If, for instance, I at this moment arise from my chair, in complete freedom, without being necessarily determined thereto by the influence of natural causes, a new series, with all its natural consequences *in infinitum*, has its absolute beginning in this event, although as regards time this event is only the continuation of a preceding series. For this resolution and act of mine do not form part of the succession of purely natural effects, and are not a mere continuation of them.[19]

The fact that I am capable of initiating 'an absolutely first beginning of a series of appearances'[20] means that I am a real agent in the world. Kant, here, takes great care that we do not misunderstand the significance of the freedom of agents in the world by interpreting it as being of a relative nature or as being derivative from and subsidiary to the originating powers of the first agent. The observation on the thesis warns us against a possible misapprehension about the freedom of agents in the world which could lead us to think that 'as a series occurring in the world can have only a relatively first beginning, being always preceded in the world by some other state of things, no absolute first beginning of a series is possible during the course of the world'. If agents exist in the world then they are real agents and their freedom (or power of absolute initiation) is not to be explained away by metaphorical or relativistic interpretations.

It should be interesting to note here the favourite analogy utilized in the correspondence between Leibniz

[19] [20] Observation on the thesis.

and Clarke for the purpose of clarifying the meaning of
the causality of agents including God's causality. Leibniz compares the will of a presumed agent to a balance
'in which everything is alike on both sides'. Then the
motives influencing the will are compared to the weights
acting on the balance. Leibniz insists that actions and
choices can issue from such a will only if there is a sufficient motive to direct it in favour of one choice rather
than another. This happens very much after the fashion
in which a weight added to one side of the balance will
necessarily tilt it in favour of that weight. The direction
of the will in its choices is externally determined by the
weight of the motives (for God these are always rational
motives) acting upon it. The stronger motive will always
determine the direction of the will in its choices. Should
a case arise where the 'weights' of the competing motives
are exactly alike then no choice could possibly occur.
Leibniz explains this point by referring to the analogy
of the balance. He says:

He [Clarke] takes it for granted, that if there be a balance,
in which everything is alike on both sides, and if equal
weights are hung on the two ends of that balance, the whole
will be at rest. 'Tis because no reason can be given, why one
side should weigh down rather than the other.[21]

For Leibniz this mechanistic model of understanding
the nature of causality remains supreme and absolutely
nothing can escape its hold and sweep. The antithesis of
the third antinomy affirms this point unambiguously.
In opposition to all this Clarke maintained that a true
agent differs from a balance in that it has its own principle of action in independence from (and at times in
opposition to) the external forces (motives, weights)

[21] Leibniz's second letter, para. 1.

affecting it. This is the free causality which the thesis of the antinomy affirms as operative in the world alongside with, and in exception to, the mechanistic causality represented by the analogy of the balance. Clarke explains this principle of action inherent in agents by describing it as

. . . the power of self motion or action: which in all animate agents, is spontaneity; and, in moral agents, is what we properly call liberty.[22]

In the observation on the thesis Kant raises a curious question. Granted the claims of the thesis about free causality one may still ask, how is this power of spontaneous origination to be 'comprehended' or 'understood'? For instance, Kant says at one point that although the power of spontaneously beginning a series

[22] Fifth letter, paras. 1–20. In commenting upon Leibniz's analogy of the balance Clarke gives a fuller explanation of his point of view on this subject. He says: 'This notion leads to universal necessity and fate, by supposing that motives have the same relation to the will of an intelligent agent, as weights have to a balance; so that of two things absolutely indifferent, an intelligent agent can no more choose either, than a balance can move itself when the weights on both sides are equal. But the difference lies here. A balance is no agent, but is merely passive and acted upon by the weights; so that, when the weights are equal, there is nothing to move it. But intelligent beings are agents; not passive, in being moved by the motives, as a balance is by weights; but they have active powers and do move themselves, sometimes upon the view of strong motives, sometimes upon weak ones, and sometimes where things are absolutely indifferent.' (Clarke's fourth letter, paras. 1 and 2.) '. . . the balance, for want of having in itself a principle or power of action, cannot move at all when the weights are equal: but a free agent, when there appear two, or more, perfectly alike reasonable ways of acting, has still within itself, by virtue of its self-motive principle, a power of acting: and it may have very strong and good reasons, not to forbear acting at all; when yet there may be no possible reason to determine one particular way of doing the things, to be better than another.' (Clarke's fifth letter, paras. 1–20.)

in time has been proved to the satisfaction of the partisans of the thesis, this does not mean that how this power is possible is thus 'understood'. In fact he affirms that this is not 'understood'.[23] Kant points out that this anomaly need not be taken against the thesis and its claims because the antithesis faces the same problem. The antithesis also produces a demonstration of its claims without rendering the principle of universal causal determinism 'comprehended' or 'understood'. Kant writes, in the observation on the thesis the following about the power of spontaneous origination:

How such a power is possible is not a question which requires to be answered in this case, any more than in regard to causality in accordance with the laws of nature. For, [as we have found], we have to remain satisfied with the *a priori* knowledge that this latter type of causality must be presupposed; we are not in the least able to comprehend how it can be possible that through one existence the existence of another is determined, and for this reason must be guided by experience alone. (A448–B476)

He repeats the same point in the observation on the antithesis. Speaking of the claim of the antithesis concerning the absence of a dynamical beginning of the world, he says:

The possibility of such an infinite derivation, without a first member to which all the rest is merely a sequel, cannot indeed, in respect of its possibility, be rendered comprehensible.

Again, Kant explains that this 'enigma in nature' need not mitigate against the claim of the antithesis because there are 'many fundamental synthetic properties and forces, which as little admit of comprehension'.[24]

[23] Observation on the thesis, A450–B478.
[24] A449–B477.

I think that Kant means to say here something more commonplace than his forbidding language seems to signify at first glance. Causality, whether of the free spontaneous type, or of the deterministic type, is not the sort of relation which can be 'understood' or 'comprehended' purely *a priori*. It cannot be derived from purely 'intelligible' principles or explained still further in terms of what Newton called 'hypotheses'.[25] If the power of spontaneous origination, and the power of one

---

[25] By 'hypotheses' Newton meant 'ideas about the world which were not deductions, through experiment, from sensible phenomena, or exactly verifiable in experience'. He wrote the following passage explaining his meaning: 'Whatever is not deduced from the phenomena is to be called an hypothesis; and hypotheses, whether metaphysical or physical, whether of occult qualities or mechanical, have no place in experimental philosophy. In this philosophy particular propositions are inferred from the phenomena, and afterwards rendered general by induction. Thus it was that the impenetrability, the mobility, and the impulsive force of bodies, and the laws of motion and of gravitation, were discovered.' In spite of his polemics against hypotheses, Newton, following the intellectual habits of his day, held that gravity, impulsive forces, etc., were only 'phenomena' that required further explanation in terms of 'deeper principles' inherent in the nature of things or in terms of more hidden and intelligible causes. As a strict scientist Newton proclaimed his ignorance as to the 'real causes' of gravitation. In this sense we would have then only a descriptive account of the force of gravity without a real 'understanding' of its nature. But in his weaker moments Newton did yield to the prevalent habits of thought and produced quite a few fanciful 'hypotheses' to explain gravitation and similar phenomena. A good example is his attempt to explain gravitation, electrical attraction, and repulsion, etc., in terms of the condensation and rarefaction of the ethereal spirits. (E. A. Burtt, *The Metaphysical Foundations of Modern Science*, Doubleday Anchor Books, Garden City, New York, 1954, pp. 215, 218, 273.) Kant's reference, in the observation on the thesis and antithesis, to: 'synthetic properties and forces which do not admit of comprehension', 'this enigma in nature', and 'our inability to comprehend how one existence determines another'; is to be understood in terms of the habits of thought of that age as briefly mentioned above.

existent to determine another existent cannot be 'understood' in this rationalistic (and *a priori*) sense of 'understanding' then many of the 'fundamental synthetic properties and forces' discussed in 'natural philosophy' would never be understood either. These forces and properties being synthetic can be neither derived from so-called purely intelligible principles *a priori* nor fruitfully referred back to 'hypotheses' (in Newton's sense). Kant emphasizes that change in the world, whether causally conceived after the fashion of the thesis or of the antithesis, cannot be 'excogitated *a priori*' but is above all a matter of experience. He says on this point:

> For were you not assured by experience that alteration actually occurs, you would never be able to excogitate *a priori* the possibility of such a ceaseless sequence of being and not being.[26]

This amounts to a reaffirmation of the general Kantian standpoint to the effect that whether there is, in the nature of things, free causality or only strict determinism cannot be dogmatically demonstrated. Kant had already explained himself on this point in the *Analytic*. He wrote:

> Had we attempted to prove these analogies dogmatically; had we, that is to say, attempted to show from concepts . . . that every event presupposes something in the preceding state upon which it follows in conformity with a rule . . . all our labour would have been wasted. (B264–A217)

Again, it should be possible for us to correlate this antinomy with the *Dissertation* of 1770 as a whole. According to 'the principles of the form of the sensible world' the idea of an infinite causal explanatory sequence

---

[26] Observation on the antithesis.

is impossible and an unconditioned causal item has to be posited, and this is the claim of the thesis. According to 'the principles of the form of the intelligible world' the idea of an infinite (never completed) causal explanatory sequence is not only legitimate but necessary. In other words, according to these latter principles we cannot admit the possibility of a causal condition which is itself unconditioned. And this is precisely the claim of the antithesis.

III

The antithesis of the third antinomy states:

There is no freedom; everything in the world takes place solely in accordance with laws of nature.

Strawson devotes about three lines to the explication of the proof of the antithesis. He sees in the antithesis and its proof a simple denial of freedom which is supported, consistently with the conclusion of the argument of the second Analogy, by an appeal to the universal application of the principle of causality.[27] I think the proof of the antithesis deserves more careful examination and analysis. There is in it more than meets the eye at first glance.

We may summarize the proof of the antithesis in the following points staying quite close to Kant's own terminology and manner of statement:

1. Assume that there is such a thing as a state(s) which possesses the 'power of absolutely beginning' another state(s_2) and by virtue of that of 'absolutely beginning' a series of consequences subsequent to s_2, as well (s_3, s_4...).

2. This means that not only the series s_2, s_3, s_4... has

[27] *The Bounds of Sense*, pp. 208–9.

an absolute beginning (with s_2), but also that the spontaneous act initiating the series ('the causality of the cause' s) is itself an absolute beginning. That is to say the spontaneous act is not determined in any way by the antecedent conditions of s and it does not follow from s according to any fixed law or pattern.

3. Now, if s is to be considered as the cause which gave rise to the series s_2, s_3, s_4... then the act of initiating the series must 'belong' to s, i.e. it must be imputable to the states of s antecedent to the initiation of s_2 and its consequences.

4. Accordingly, we face the following dilemma: either (a) s is not the cause of s_2, s_3, s_4... since the spontaneous act of initiating the series is in no way imputable to the antecedent conditions of s and in no way follows from them according to a fixed pattern or rule; or (b) the act of initiating the series is imputable to s on the grounds that s is the cause of the series. In this latter case the act of initiating the series would be neither 'spontaneous' nor 'free' since it would then follow from the antecedent states of s according to a fixed rule.

5. It becomes redundant to point out that (a) contradicts the initial assumption which regards s as the cause 'to which belongs' the power of absolutely beginning the series in question, while (b) amounts to a denial of freedom in the course of events since 'if freedom were determined in accordance with laws, it would not be freedom, it would simply be nature under another name'.[28]

It should be clear that the crux of the argument of the antithesis (against the assumption of the truth of the thesis) is to raise a certain basic question about the

[28] Proof of the antithesis.

nature of the relationship holding between the supposed
spontaneous act of origination on the one hand and the
agent presumably 'responsible' for that act on the other.
The purpose of the argument of the antithesis is to show
that in whichever way we try to construe this relationship we are landed in grave difficulties and absurdities.
This has the effect of discrediting the initial assumption
which leads to these difficulties. The proof, in other
words, is concerned with pointing out the 'insurmount-
able difficulties in the way of admitting any such type
of unconditioned causality'.[29] These difficulties surrounding transcendental freedom pertain to the problem
of the proper grounds to which a supposed spontaneous (free) action is to be imputed.[30] If according to
the thesis absolute spontaneity is the only proper ground
to which a free action may be imputed then the question arises as to the sense in which a free action may be
said to 'belong' to the supposed author (agent) of that
action.

If the presumed free action really 'belongs' to an agent
and is connected with its states in a definite and specifiable manner, then its spontaneity is lost. In such a case
we simply have the old causal connection between an
action and the antecedent states of the agent but called
by another name. If, on the other hand, we affirm that
the free action is *sui generis* in the sense that it is
imputable only to absolute spontaneity and not to the
antecedent states of an agent, then we would create a
gap between actions and agents such that the former
cannot be said to be referable, in any definite sense, to
the latter. This will have the effect of abolishing 'real
agents' and 'real choices' from the world without which

[29] [30] Observation on the thesis.

the thesis cannot stand. We will have on our hands presumed 'agents' that cannot be considered the authors of choices in any specifiable sense of the term. We will also have presumed 'choices' which are not the choices of any specific agent since such choices are imputable to pure spontaneity only and not to the antecedent states of the agent itself.

The sense of these criticisms of the idea of the freedom of agents (as conceived in the thesis) was already urged by Leibniz against Clarke. The famous German philosopher pointed out to Clarke that his notion of spontaneity would simply make nonsense out of the whole idea of choice. He wrote:

In things absolutely indifferent, there is no [foundation for] choice, and consequently no election, nor will; since choice must be founded on some reason, or principle.[31]

Leibniz insisted, furthermore, that on Clarke's conception, choice would be simply blind chance and thus no choice at all.[32] For him:

A will without reason, would be the chance of the Epicureans. A God, who should act by such a will, would be a God only in name.[33]

In another passage Leibniz points out that Clarke's conception of freedom creates a division between the agent and his actions by artificially separating the former from the motives proper to him and out of which voluntary choices flow.[34] And I have tried to show that this is a central point in the argument of the antithesis. Leibniz, then, reaffirms his deterministic conception according to

[31] Fourth letter, para. 1.
[32] Fifth letter, para. 7. See also para. 70.
[33] Fourth letter, para. 18.
[34] Fifth letter, para. 15.

which every state of the agent is causally determined by its antecedent states.[35]

If the line of thought represented in the thesis is pushed to its logical conclusions, the entire course of events would be utterly atomized. Kant brings out this point clearly when he writes the following about the idea of freedom proposed by the thesis: '. . . the kind of connection which it assumes as holding between the successive states of the active causes renders all unity of experience impossible'.[36] The free causality introduced by the thesis is 'blind and abrogates those rules through which alone a completely coherent experience is possible'.[37]

The argument of the antithesis derives part of its strength from the fact that the partisans of the thesis are also committed (as Newtonians) to the idea of a 'law abiding natural order' and to the notion of a completely coherent nature and experience. Clarke agrees wholeheartedly with the principle 'that nothing happens without a sufficient reason (cause), why it should be so, rather than otherwise'. The argument of the antithesis works to point out that the partisans of the thesis contradict this initial commitment to an orderly nature when they admit the reality of spontaneous origination. From this point of view the claims of the thesis about real agents, who have the power of spontaneous origination, amount to an admission of actual exceptions to the sweep of the principle of sufficient reason. This would make a mockery out of the idea of an orderly nature and of a completely coherent experience. We should note again that Leibniz directed this sort of criticism against Clarke's ideas. For

[35] See, for example, his fifth letter, para. 91.
[36] [37] Proof of the antithesis.

example, he wrote: 'The author (Clarke) grants me this important principle; that nothing happens without a sufficient reason, why it should be so, rather than otherwise. But he grants it only in words, and in reality denies it. Which shows that he does not fully perceive the strength of it.'[38] Of course, to grant such an important principle in words only is to rob actuality of all order and coherence.

The argument of the antithesis espouses Leibniz's very strict interpretation and application of the law of sufficient reason, and uses it to combat the opposing point of view. For example, Leibniz's insistence that each and every occurrence, not excepting God's choices, must be referred to a set of explaining conditions (the sufficient reason of the occurrence) is really identical with the central deterministic claim of the antithesis. Leibniz stated this point very clearly in a passage quoted earlier:

For the nature of things requires, that every event should have beforehand its proper conditions, requisites, and dispositions, the existence whereof makes the sufficient reason of such an event.[39]

The paragraph coming right after in Leibniz's letter makes clear that God's nature is no exception to this rule as Clarke would have us think. The antithesis also is against the idea of a first dynamical beginning of the universe which is not subject to the universal causal principle.[40]

[38] [39] Fifth letter, para. 18.

[40] This is why Kant says that the supporters of the antithesis show 'a perfect uniformity in manner of thinking and complete unity of maxims . . .' (A465–B493). In contrast, he describes the partisans of the theses as breaking 'the thread of physical inquiries' at the point which suits their convenience (A470–B498).

In the final paragraph of the observation on the anti-thesis Kant brings out two specific objections against the thesis. The first is, I think, directed against the idea of a first free agent (first cause), the second against the idea of free agents which are part of the processes of the world. It seems to me that Kant is trying to make the following point in this passage which is presented from the point of view of the antithesis. If the partisans of the thesis agree that causally determined change cannot be 'excogitated *a priori*' but is known only through experience, then their assumption of a first agent becomes a very bold one, to say the least. After all such an agent would have to be above and beyond all experience which automatically puts him outside the scope of human knowledge. Kant states his point in the following way:

Even if a transcendental power of freedom be allowed, as supplying a beginning of happenings in the world, this power would in any case have to be outside the world (though any such assumption that over and above the sum of all possible intuitions there exists an object which cannot be given in any possible perception, is still a very bold one).[41]

Experience lends no authority to the claims of those who invent 'an absolutely first state of the world, and . . . an absolute beginning of the ever flowing series of appearances . . . (in order to procure) a resting place for (their) imagination by setting bounds to limitless nature'.[42] On the contrary, experience seems to indicate in the opposite direction:

Since the substances in the world have always existed—

[41] Observation on the antithesis.
[42] Observation on the antithesis, first paragraph.

at least the unity of experience renders necessary such a supposition—there is no difficulty in assuming that change of their states, that is, a series of their alterations, has likewise always existed, and therefore that a first beginning, whether mathematical or dynamical, is not to be looked for.[43]

Concerning the claims made by the thesis about the actuality of real free (spontaneous) agents in the world Kant points out that this sort of admission is even less excusable, from the point of view of what experience warrants, than the admission of a first agent and source of the world process. The presence of spontaneity in the processes of the world would simply reduce nature to total disorder and lawlessness, thus rendering the very notion of 'experience' itself impossible and senseless. Kant expounds this point at length in the last parts of the observation on the antithesis. He says:

But to ascribe to substances in the world itself such a power, can never be permissible; for, should this be done, that connection of appearances determining one another with necessity according to universal laws, which we entitle nature, and with it the criterion of empirical truth, whereby experience is distinguished from dreaming, would almost entirely disappear. Side by side with such a lawless faculty of freedom, nature [as an ordered system] is hardly thinkable; the influences of the former would so unceasingly alter the laws of the latter that the appearances which in their natural course are regular and uniform would be reduced to disorder and incoherence.[44]

Now, since transcendental freedom 'is not to be met in any experience' it is therefore an empty 'thought-

[43] Observation on the antithesis, first paragraph.
[44] Observation on the antithesis.

entity'.[45] Furthermore, as long as the partisans of the thesis and antithesis insist, each in turn, that the system of metaphysical ideas he dogmatically stands for is a true description of the nature of things as they are in themselves, the conflict will for ever remain without a genuine solution.

[45] Proof of the antithesis.

THE FOURTH ANTINOMY

THE last of the antinomies deals with the problem of a necessary being. The thesis holds that 'there belongs to the world, either as its part or as its cause, a being that is absolutely necessary'. The proof of the thesis takes great pains to show that if such a necessary being exists then it is an integral part of the phenomenal world and cannot be separated from it. The antithesis holds that 'an absolutely necessary being nowhere exists in the world, nor does it exist outside the world as its cause'.

Before I proceed to the examination of the thesis and its proof I would like to refer to the tendency among many Kantian commentators to regard the fourth antinomy as dealing basically with a theological problem. One of the main points upon which they dwell in their treatment is the traditional 'first cause argument' (cosmological argument). They find in the proof of the thesis an argument always associated, in the history of philosophy, with the demonstration of the existence of God. In fact Norman Kemp Smith tends to collapse the third and fourth antinomies together on account of the prominence of the traditional first cause argument in both antinomies. As a result, his discussion gives the last two antinomies a definitely theological content. He writes at one point: 'Thus Kant's proof of freedom in the thesis of the third antinomy is merely a corollary from his proof of the existence of a cosmological or theological unconditioned. . . .'[1] Let us simply note, at this point,

[1] *Commentary*, p. 497.

Smith's use of 'a cosmological unconditioned' inter-
changeably with 'a theological unconditioned'.[2]

Weldon comments on the subject of the fourth antinomy only in the context of his discussion of Kant's
refutation of the traditional proofs of the existence of
God.[3] For Weldon the cosmological unconditioned,
argued for in the thesis of the fourth antinomy, is not
any different from the theological unconditioned treated
in the well-known traditional theological arguments.
Similarly, Martin mentions the fourth antinomy only in
the context of his discussion of Kant's theology.[4] Ewing
also discusses this antinomy along similar lines.[5]

I would like to emphasize that, strictly speaking,
Kant's concern in the fourth antinomy is with a cosmological and not a theological problem. The traditional
theological arguments deal with a theological unconditioned which is definitely not a part of the phenomenal
world but is separate from it. In contrast to this the
arguments of the fourth antinomy (both in the thesis
and antithesis) stress again and again that the necessary
being under consideration, if it exists, is a part of the
phenomenal world and cannot be separate from it. The
main concern, then, is with a cosmological and not a
theological unconditioned. This is made clear in more

---

[2] For the modern reader the distinction between theology and
cosmology might seem trivial and irrelevant. However, this was not
true in the eyes of the thinkers of Kant's century. Such a distinction
still meant very much for them. It is not a historically insignificant
fact that Kant dealt with the 'System of Cosmological Ideals' in
separation from 'the Ideal of Pure Reason' where the theological
idea is critically examined in the *First Critique*.

[3] *Kant's Critique of Pure Reason*, p. 230.

[4] *Kant's Metaphysics and Theory of Science*, p. 165.

[5] A. C. Ewing, *A Short Commentary on Kant's Critique of Pure
Reason*, pp. 218, 219.

than one passage in the observation on the thesis. For instance, Kant unambiguously states about the necessary being in question:

This cause, even if it be viewed as absolutely necessary, must be such as can be thus met with in time, and must belong to the series of appearances.

Since the antinomy is concerned with a cosmological as distinguished from a theological unconditioned, Kant explicitly rejects any use of the ontological argument (which, according to Kant, underlies all proofs for the existence of God) in the antinomy and relies only on the cosmological mode of arguing.[6] He explains that the thesis utilizes an impure form of the cosmological argument. The pure form does not settle, strictly speaking, the question of whether the necessary being is a part of the world or distinct from it. The impure form utilized in the thesis is adopted to settle this very question, i.e. to lead to the final conclusion that the necessary being which exists is an integral part of the world and not otherwise.[7] To establish the opposite: '. . . we should require principles which are no longer cosmological and do not continue in the series of appearances'.[8] These required principles, Kant affirms, belong to a *transcendent* philosophy (as distinguished from cosmology) which 'we are not yet in a position to discuss'.[9]

This theological bias in looking at the fourth antinomy leads Caird to hold that Kant's repeated insistence, in the proof of the thesis, that the necessary being must be in the world is simply an 'irregularity'. Accordingly, when the irregularity is overlooked the point of the argument in the thesis would be 'that there must be a

[6] [7] [8] [9] Observation on the thesis.

necessary being either *in* or *out* of the world'.[10] This conclusion on the part of Caird contradicts the plain sense of the texts. In another context Kant provides us with a statement of the conflict in theologically neutral terms. He formulates the conflict thus:

. . . whether there exists any being completely unconditioned and necessary in itself, or whether everything is conditioned in its existence and therefore dependent on external things and itself contingent. (A481–B509)

All this does not mean, however, that the cosmological conflict about the necessary being does not have important consequences for theology. Again, it does not mean that the conflict does not spring from a theological background and does not have definite theological overtones. In fact Kant devotes some of his efforts to a discussion of the theological implications of the fourth cosmological conflict. He does this in the context of his investigation of the extra-rational motives and reasons which often lead people to side with the thesis or the antithesis as the case may be. Kant says, in this connection, that we find on the side of the thesis 'a certain *practical interest* in which every well-disposed man . . . heartily shares'. This *practical interest* consists in the belief: 'that all order in the things constituting the world is due to a primordial being, from which everything derives its unity and purposive connection—these are so many foundation stones of morals and religion' (A466–B494).

Partly, it is this predominantly theological (and practical) manner of regarding the antinomy which leads Paulsen to describe the thesis as being part of the 'well-meaning' philosophy which works in the interests of

[10] *The Critical Philosophy of Kant*, vol. ii, p. 49.

theology by demonstrating the existence of God as a necessary being. In the same vein Paulsen describes the antithesis as representative of 'the empiristic and materialistic mode of thought, which was based on the natural sciences' and which contested the 'good' claims of the thesis.[11]

But Paulsen is completely mistaken on this matter. It is the thesis which is closer to the 'empiristic' and materialistic 'modes of thought' as practised by the eighteenth-century scientists and not the antithesis. After all, the proof of the thesis stresses over and over that the necessary being demonstrated is an integral part of the material (phenomenal) world in space and time. As pointed out in the first chapter it is the Galilean-Newtonian conception of the universe that has supplied the basis of modern materialism. As we shall see, the claim of the thesis that the necessary being is part of the world in space and time is simply a restatement of a conclusion which follows from the Newtonian view of nature and its premises. This conclusion was very clearly discerned by Leibniz and pointed out to Clarke many times throughout the correspondence. For instance, Leibniz wrote: 'Natural religion itself, seems to decay (in England) very much. Many will have human souls to be material: others make God himself a corporeal being.'[12] Again he wrote to Clarke warning him that according to the views he is defending 'God will be part of nature'.[13] Accordingly, he accuses Clarke of falling into the 'errors' of the materialists and specifically of Spinoza.[14]

[11] F. Paulsen, *Immanuel Kant*, pp. 215–16.
[12] First letter, para. 1.
[13] Fifth letter, para. 111.
[14] Second letter, para. 7.

In other words, the thesis is very far from working in the interests of theology and the 'well-meaning' philosophy. It is working, in the final analysis, in the interests of the materialistic point of view which prevailed in scientific circles at that time. Leibniz saw this point very well and used it effectively to embarrass his opponents. The source of the embarrassment for Clarke was a certain ambiguity in the expression of his theological position. He often spoke, in the correspondence, about God in such language as:

In all void space, God is certainly present.[15]

God sees all things, by his immediate presence to them.[16]

God, being omnipresent, is really present to everything, essentially and substantially.[17]

[He is] substantially present everywhere.[18]

In expressing himself in this fashion Clarke is being a faithful Newtonian. But then if God is substantially present 'everywhere' and 'everywhen'; and if 'where' and 'when' can mean, for Newtonians, only a portion of absolute space and a segment of absolute time, it would seem to follow that God is corporeal (He occupies space and endures in time). Leibniz correctly concluded from the Newtonian assumptions that God is a part of nature or the material world, according to them. Newton himself held that:

Since every particle in space is always, and every indivisible moment of duration is everywhere, certainly the

[15] Fourth letter, para. 9.
[16] First letter, para. 3.
[17] Third letter, para. 12.
[18] Ibid., para. 15.

Maker and Lord of all things cannot be never and no-where.[19]

Furthermore, God's presence in all container-space and all infinite receptacle-time has two characteristics according to Newton:

He is omnipresent not virtually only, but also substantially; for virtue cannot subsist without substance . . . God suffers nothing from the motion of bodies; bodies find no resistance from the omnipresence of God.[20]

No wonder Leibniz compared the views of Clarke (and of course Newton) to Spinoza's.

In addition to being a faithful Newtonian, Clarke was also a good Christian. He, therefore, hated the charge of 'materialism' and resisted the conclusion that God is corporeal and a part of nature. In deference to his religious beliefs he wanted to hold a more orthodox formal theism where God is the Creator, Lord, and Governor of the world and not a part of it. On account of this ambiguous position Leibniz described him as a Christian materialist.[21] When Leibniz attempted to force Clarke into making a clear-cut choice between the conception of God as *intelligentia mundana* and as *intelligentia supra mundana*, Clarke escaped the dilemma by saying that He is neither and He is also both. Clarke covered the ambiguity of his position by appealing to

[19] 'Extracts from Newton's *Principia*', *The Leibniz–Clarke Correspondence*, p. 167.

[20] 'Extracts', p. 168. Admittedly if God occupies absolute space then He has to occupy it in a peculiar manner, i.e. without filling it the way ordinary bodies do. This concession is made imperative by the traditional theological beliefs of Newton, Clarke, and many others like them.

[21] Second letter, para. 1.

traditional rhetorical and paradoxical religious language.
He wrote:

God is neither a *mundane intelligence*, nor a *supra-mundane intelligence*; but an omnipresent intelligence, both in and without the world. He is in all, and through all, as well as above all.[22]

The point I would like to argue later is that the thesis and its proof are basically a restatement of the idea implicit in the Newtonian cosmology, viz. that the necessary being (if it exists) is a part of the material world in absolute space and time.

Contrary to what Paulsen holds the claim of the antithesis does not represent 'the empiristic and materialistic mode of thought, which was based on the natural sciences'. The antithesis denies that a necessary being exists either as a part of the world or as distinct from it. The claim of the antithesis is a natural result of the particular way in which Leibnizian metaphysics viewed the nature of the universe. Clarke correctly pointed out to Leibniz that granted his (Leibniz's) metaphysical assumptions then there would be no God at all of the universe. God, according to Clarke, would be no God at all on the Leibnizian point of view. Following is the plain atheistic result to which the Leibnizian position leads, as formulated by Clarke:

. . . a sceptic will easily argue still farther backwards, and suppose that things have from eternity gone on (as they now do) without any true creation or orginial author at all, but only what such arguers call all-wise and eternal nature.[23]

Clarke thinks that God is part of His Kingdom and

[22] Second letter, para. 10.
[23] First letter, para. 4.

administers it. He accuses Leibniz of holding views
which lead to the exclusion of God from His Kingdom.
He says:

> The notion of the world's being a great machine, going
> on without the interposition of God, as a clock continues to
> go without the assistance of a clockmaker; is the notion of
> materialism and fate, and tends, (under pretence of making
> God a *supra-mundane intelligence*,) to exclude providence
> and God's government in reality out of the world.[24]

Again he attacks Leibniz's typical arguments in the
correspondence on the grounds that they lead to a
denial of the existence of God. For instance Clarke
writes:

> This argument, if it were good, would prove that whatever
> God can do, he cannot but do; and consequently that he
> cannot but make every thing infinite and every thing
> eternal. Which is making him no governor at all, but a
> mere necessary agent, that is, indeed no agent at all, but
> mere fate and nature and necessity.[25]

Clarke realized that the eternalistic metaphysics of
his opponent had no genuine use for God (or a necessary being) in any normal theistic sense. He used this
point to embarrass Leibniz. Clarke correctly pointed out
that granted the spatial and temporal infinity of the
universe there is no need for a God in the sense of the
initiator (creator), sustainer, and ruler of the world.
Leibniz disliked the charge of atheism (he always
thought of himself as a good Christian too) and tried to
extricate himself from the difficulty. His attempt consisted of a little intellectual trick: although the world is
spatially infinite we may regard it as temporally finite

24 First letter, para. 4.
25 Fourth letter, paras. 22 and 23.

a parte ante only (and not *a parte post*), thus making room for a possible first cause of the universe.[26] The trick was so obvious and arbitrary that Leibniz did not insist on it and it was not taken seriously in the correspondence. Clarke replied by reaffirming the more consistent position: '. . . and consequently the material universe must be not only boundless, but eternal also, both *a parte ante* and *a parte post*, necessarily and independently on the will of God'.[27] Later on Kant himself wrote the following disapproving observation, on Leibniz's little trick, in the *Dissertation* of 1770:

But if a simultaneous infinite be admitted, the totality of the successive infinite must also be conceded; while if the latter is denied, the former must also be given up. (42–3)

Leibniz cannot have his cake and eat it too.

In this light we may regard the claim of the antithesis to imply:

1. A rejection of Clarke's ambiguous view of God; as being both in and yet above the world.

2. An affirmation of the 'atheism' implicit in the eternalistic metaphysics of Leibniz.

II

Kant leaves no doubt in the fourth antinomy that what he means by 'the world' is 'the sum-total of all appearances' in uniform space and time. His concern is with the sensible material universe. For reasons of convenience the discussion in the antinomy neglects space and utilizes the language and idiom of time to formulate

---

[26] Fifth letter, para. 74.
[27] Fifth letter, paras. 73–5.

problems and present arguments. Kant says the following about the proof of the thesis:

The former argument [i.e. of the thesis] takes account only of *the absolute totality* of the series of conditions determining each other in time, and so reaches what is unconditioned and necessary.[28]

The 'world' for the thesis, is, then, an absolute totality of appearances (a synthetic totality) which is also finite. The necessary being, if it exists, has to be part of this totality and subject to its universal conditions. Kant emphasizes this truth about the thesis and its assumptions in the following manner:

. . . the highest condition or cause can bring the regress to a close only in accordance with the laws of sensibility, (i.e. space and time) and therefore only in so far as it itself belongs to the temporal series.[29]

The proof of the thesis consists really of two separate proofs. The first is devoted to showing that the world does have an absolutely necessary cause. I shall call it (P_1). The second proof is designed to demonstrate that this absolutely necessary being is a part of the world. I shall refer to it as (P_2). The argument of P_1 may be summarized in the following points:

1. Assume that the world has no absolutely necessary (unconditioned) cause.

2. The sensible world as a whole of appearances contains a series of alterations in time.

3. Every such alteration 'stands under its (causal) condition which precedes it in time and renders it necessary'.

4. Accordingly, every such conditioned alteration presupposes a series of preceding conditions which form the

[28] Observation on the antithesis. Kant's italics.
[29] Observation on the thesis.

complete and sufficient causal determinants of the existence of the alteration in question.

5. Now, if this series of preceding conditions were to go on indefinitely then the series would always remain incomplete and no truly complete and sufficient causal explanation of the alteration would have been produced.

6. Since every conditioned alteration presupposes, 'in respect of its existence, a complete series of conditions' it follows that the series of preceding conditions cannot remain for ever incomplete. It ends with a condition which is not itself conditioned by anything preceding it.

7. This unconditioned condition is the absolutely necessary being which is the cause of the world.

We should observe here that P_1, like the proof of the thesis in the third antinomy, rests on a certain interpretation of the meaning of the completeness of the regressing series of causal conditions. This meaning implies that the enumeration of the members of the regressive series occupies (in principle) a finite stretch of time. Otherwise, it would be impossible to comprehend the world of phenomena in a simple 'whole' or complete 'totality'. The absence of such a totality would render all causally conditioned alterations in the world inconceivable considering that no sufficient and complete explanation for them could be ever given in that case.

Accordingly, the crux of the argument in P_1, may be restated as follows:

(a) On the assumption that the regressive series of causal conditions does not terminate in an unconditioned condition the world would not form a whole or a totality.

(*b*) The world is an absolute finite totality of phenomena.[30]

(*c*) Therefore: the series of causal conditions does terminate in an unconditioned condition.

Obviously, without granting (*b*), which is part and parcel of the Newtonian cosmology, the argument will not work. Kant is quite aware of this. For he says that the argument of the thesis 'takes account only of *the absolute totality* of the series of conditions determining each other in time, and so reaches what is unconditioned and necessary'.[31] Furthermore, the proof that a necessary being exists reflects the claims of Clarke (on the level of the cosmological ideas) that without such a being the actual finite world would be inconceivable; and that the assumption of the infinitude of the world leads to a denial of the reality of such a being. It renders His existence only nominal. Clarke explains himself on these points in many passages of the correspondence. The following statement will suffice by way of illustration:

But if, on the contrary, the material universe cannot be finite and moveable, and space cannot be independent upon matter; then (I say) it follows evidently, that God neither can nor ever could set bounds to matter; and consequently the material universe must be not only boundless, but eternal also, both *a parte ante* and *a parte post*, necessarily and independently on the will of God.[32]

There remains one further observation to be made about P_1. This proof is really more than a simple employ-

---

[30] The antithesis, according to Kant, implies 'the existence of a chain of causes which in the regress of their conditions allow of no *absolute totality*'. While the thesis implies that the chain forms precisely such a *totality* (A452–B570).

[31] Observation on the antithesis. Kant's italics.

[32] Fifth letter, paras. 73–5.

ment of the traditional first-cause argument. It is an application of Kant's notion of 'analysis' as explained in the *Dissertation* of 1770. Analysis in this sense is 'a regress from consequence to ground' which 'rests on conditions of time' (36). Such an analysis will not yield a 'necessary first ground' unless the process 'can be brought to a conclusion in a finite and assignable time'. Always central to the argument is the 'scientific' assumption of the Newtonians that the material universe is finite in space and time. Accordingly, P_1 would be arguing that if a series of conditioned (contingent) changes is 'given' then it can be shown that a necessary and unconditioned item of the series is also given. This is exactly analogous to the proofs of the theses of earlier antinomies which argued that if a complex of substances is given then it can be shown 'that simples, and a world, are also given' (38). All these conclusions are no more than philosophical restatements of the then prevalent Newtonian view of the nature and constitution of the material world.

Granted that a necessary being exists as the first cause of the world the question arises as to the relationship of this being to the material world itself. P_2 is devoted to proving that this necessary being is an integral part of the sensible world taken as the totality of appearances.

Following is a statement and explanation of the contents of P_2:

1. The necessary being cannot be considered as the beginning cause of the series of alterations forming the sensible world in time unless it is itself regarded as part of the sensible world and in time.

2. This is so for the following reason: by definition,

to be a beginning is to precede in time what is begun, i.e. to exist at a moment of absolute time which precedes the moment at which 'the thing begun' came into existence.

3. If the necessary being is the beginning cause of the sensible world (as was demonstrated earlier) then it existed at a moment of absolute time which preceded the moment at which the sensible world came into existence.

4. Therefore: 'the cause itself, must belong to time and so to appearance . . . ', i.e. it 'cannot be thought apart from that sum of all appearances which constitutes the world of sense' because, then, it would not be the beginning cause of the world in any recognizable sense of the term 'beginning'.

5. The final conclusion of the proof is:

Something absolutely necessary is therefore contained in the world itself, whether this something be the whole series of alterations in the world or a part of the series.

Some analysis of the final conclusion is needed here. It seems to be saying that 'something absolutely necessary is contained in the world' and that this can have two senses: (a) that the series of alterations forming the world is 'necessary' as a whole; (b) that at least one item of the series is necessary and the remaining items are causally dependent upon it. By virtue of this causal dependence the other items may also be considered as necessary.

According to the thesis 'necessity' means 'not causally dependent on anything else'. X is necessary if it is existentially self-sufficient (*causa sui*). Kant explains in the *Dissertation* that in this sense 'X is necessary' means that 'its existence is securely established, apart from all dependence' for such dependence 'is not appropriate to things necessary' (67). In contrast to this we have a different interpretation of the meaning of 'necessity' in the antithesis. The antithesis is predicated on the idea that 'necessity' is no more than causal dependence. 'X is necessary' if it follows 'necessarily' from a previous condition which is its cause. Necessity, in this sense is the same as strict causal contingency (dependence).

Now we return to the final conclusion of P_2. Part (a) of the conclusion holds that the series of alterations forming the world is 'necessary' as a whole. This means that the material world taken as a complete and finite whole does not derive its existence from anything outside itself. In fact if we consider the world as an absolute totality in space and time then it would be absurd to speak about anything external to this totality. The 'world is necessary as a whole' because 'its existence is securely established, apart from all dependence'. This is a meaningful description of the world in the sense in which Newtonians find it meaningful to speak about the existence of a single material body (or the entire material universe) in infinite space and time. In effect, this is the substance of the claim of part (a) of the final conclusion of P_2.

Another way of thinking about the claim that 'the world is necessary as a whole' is in terms of saying that 'the principle of necessity' is inseparable from this totality or whole. This is part (b) of the final conclusion of P_2. It claims that the 'something absolutely necessary' contained in the world is only an item of the world series of alterations (and not the whole of it). Again, this manner of regarding the necessary being is based on the specific sense of 'necessity' explained above. This first item of the series is self-sufficient in its existence while all the other items are causally traceable (dependent) to it in a finite amount of time. In this sense too the world may be regarded as a causally coherent and necessary finite whole.

Kant realized, in the *Dissertation*, that to have a world in this sense 'the necessary being' has to be one of its integral and inseparable parts. In discussing the factors which enter into the definition of a 'world' he affirms the need for a principle (a form) which is 'immutable, and not liable to change' (40). Without this principle we would simply have an aggregation of things and not a world. Kant makes his point as follows:

Thus in any world there is given a certain form which is to be reckoned as belonging to its nature, constant, in variable, as the eternal principle of every contingent transitory form pertaining to the state of the world. (41)

Concerning the argument of P_2 we should note that it turns on the definition of the word 'beginning'. The argument makes sense only if we understand it against the background of the concept of infinite receptable-time where absolute beginnings, ends, and precedences form real facts of nature. If we take Kant's use of 'beginning' in P_2 in a more ordinary sense (and hence a more relativistic sense) the argument will not be meaningful. In other words, there is one 'recognizable' and 'true' meaning of 'beginning', 'end', and 'precedence', for the thesis and it is definable in terms of the absolutely equable flow of the moments of time.

The insistence we find in P_2 on the idea that the necessary being is in time and an integral part of the world is again a reflection of the natural conclusion that the

Newtonian world picture leads to concerning the presence of God. If this picture is accepted in its full vigour then God would have to be in space and time, i.e. a part of the world. In the *Dissertation* Kant says the following about the Newtonians:

> They imagine the presence of God to be local, and involve God in the world as if he were comprised in infinite space all at once, although they themselves have to make up for this limitation by conceiving the locality as it were *per eminentiam*, namely, as infinite. (77)

This comment on the theological implications of the Newtonian cosmology is partly based on Clarke's view that space and time are properties of God. 'And without them, his eternity and ubiquity (or omnipresence) would be taken away.'[33] Clarke writes:

> God, being omnipresent, is really present to everything, essentially and substantially. This presence manifests itself indeed by its operation, but it could not operate if it was not there.[34]

Clarke will accept the description of God as *intelligentia supramundana* only if it does not mean to deny God's substantial presence in the world. He says:

> The phrase, *intelligentia supramundana*, may well be allowed, as it is here explained: but without this explication, the expression is very apt to lead to a wrong notion, as if God was not really and substantially present everywhere.[35]

Now, given the Newtonian doctrine that all presence is presence in absolute space and time the conclusion that God's substantial presence to all things is necessarily in space and time becomes inescapable. Hence the

[33] Clarke's fourth letter, para. 10.
[34] Third letter, para. 12.
[35] Ibid., para. 15.

Leibnizian charge of 'materialism' *à la* Spinoza against the Newtonians. In fact Clarke himself almost admitted as much:

God perceives everything, not by means of any organ, but by being himself actually present everywhere. This everywhere therefore, or universal space, is the place of his perception.[36]

Following this line of thought Clarke relativizes the distinction between the 'natural' and the 'supernatural' by rejecting the idea of an absolute difference between the two realms. He says:

But the truth is; *natural* and *supernatural* are nothing at all different with regard to God, but distinctions merely in our conceptions of things.[37]

Kant makes the following comment, in the *Dissertation*, about Clarke's (or the Newtonians) over-all views:

As regards time, philosophers involve themselves in an inextricable maze, for not only do they disconnect it from the laws of sensitive apprehension, they likewise carry it beyond the confines of the world, to the extramundane being itself, as a condition of the existence of such being. Hence the absurd questions with which they torment their minds, e.g., why God did not fashion the world many ages earlier.[38] They persuade themselves that it can easily be conceived how God perceives things present, that is, the actual things of the time in which He is; but how He foresees the future, that is, the actual things of a time in which He is not yet, they think difficult of understanding. (78)

The claims made in P_2 are not much more, then, than a restatement, in a certain kind of philosophical language, of the position to which the Newtonians are

[36] Fifth letter, paras. 79–82.
[37] Second letter, para. 12.
[38] This point was debated at some length by Clarke and Leibniz.

driven concerning the existence of the necessary being. Furthermore, Kant had already discussed this position in the *Dissertation* of 1770.

III

The proof of the antithesis consists of two parts also. The first part demonstrates that 'an absolutely necessary being nowhere exists in the world'. I shall call this part (P₃). The second part demonstrates that an absolutely necessary being nowhere exists 'outside the world as its cause'. This I shall call (P₄). The antithesis denies the Newtonian conception of the world as requiring a necessary being whether considered as a part of the world or as separate from it.

The argument of P₃ may be summarized and explained as follows:

1. Assume that the world is necessary.

2. According to the final conclusion of P₂ the necessity of the world may be understood in either of two ways:

(a) that the series of alterations forming the world has an absolutely necessary beginning which is uncaused;

(b) that although every part of such a series is causally contingent on every other part none the less the entire series taken as a whole is absolutely necessary and unconditioned.

3. (a) Violates the law of causal dependence in the world of phenomena by affirming the existence of an item which is causally not dependent on anything else.

(*b*) Is absurd since a series of alterations whose parts are all contingent cannot be regarded as a 'necessary whole' when not one of its parts is necessary.

4. Therefore: in the world there does not exist any necessary being.

Strawson considers the refutation of (*b*) on the grounds that 'a series cannot be necessary if no single member of it is necessary' to be very poor and confused.[39] However, if we put the argument in its proper context it would make more sense than one would suspect at first. The point of the refutation of (*b*) may be explicated in the following way:

1. Assume that the following series of alterations PQR . . . form a world (a totality or whole as understood by the thesis).

2. If every member of the series were necessary then PQR . . . could not form a world or a whole because: 'A whole of necessary substances is impossible. Since for each its own existence is securely established, apart from all dependence (which clearly is not appropriate to things necessary) on anything else . . .'. (67)

3. If every member of the series were contingent, then again PQR . . . could not form a world (or a whole) because a series of merely and indefinitely contingent alterations cannot form a world (by the Newtonian definition of the world).

4. It follows that PQR . . . form a world if and only if they are neither all necessary nor all contingent, i.e. only if one item among them is necessary.

[39] *The Bounds of Sense*, p. 210.

5. But (*b*) contradicts this conclusion by claiming that the world is a necessary whole of merely contingent parts.

6. Hence (*b*) is disqualified.

This proof tries to show that on the assumption of one claim of the thesis an absurdity will result. This renders the claim of the thesis untenable.

The objection in P_3 to (*a*), i.e. to the assumption that the series of alterations forming the world has an un-caused necessary beginning, is already familiar. It is the often-repeated charge directed by Leibniz against Clarke of being unfaithful to the strict principle of causal dependence. On the one hand, Clarke (and the partisans of the thesis) claims to accept this universal principle but, on the other, he proceeds arbitrarily to set up a privileged instance which is not subject to the principle itself. This is the necessary uncaused being.

Clearly the argument of the antithesis in P_3 is predi-cated on a rejection of Clarke's notion of 'necessity'. It affirms instead the Leibnizian relational meaning of necessity. On the latter view an item may be sensibly regarded as 'necessary' only relatively to another item such that the existence of the former is definitely dependent on the latter in a specified sense. Since this is the only sense of 'necessity' which the Leibnizian point of view will admit it follows that the series of alterations in the world cannot terminate in an item which is itself not necessary in the relational sense. And this is precisely the substance of the claim of P_3.

Similarly, P_3 is predicated on a rejection of the New-tonian definition of the 'world' as a necessarily finite whole. It substitutes for it the eternalistic conception of reality characteristic of Leibnizian metaphysics. P_3 achieves this rejection by showing that the Newtonian notions of 'absolute (non-relational) necessity' and of the 'world as a whole', when carefully examined lead to absurdities and arbitrary exceptions to the principle of causality (sufficient reason) as understood by Leibniz. This is sufficient to discredit the thesis and its claims and to establish the opposing claims of the antithesis.

Kant comments on this antinomial conflict in the following words:

A strange situation is disclosed in this antinomy. From the same ground on which, in the thesis, the existence of an original being was inferred, its non-existence is inferred in the antithesis, and this with equal stringency.[40]

The 'same ground' referred to in this passage is of course the principle of causal dependence to which the partisans of the thesis and of the antithesis agree in principle. But the conflict arises because the partisans of the thesis interpret causal necessity in an absolutistic manner to mean causal self-sufficiency. Then they proceed to draw their conclusions from this interpretation with perfect stringency. On the other hand, the partisans of the anti-thesis follow the strategy of (i) assuming the thesis; (ii) interpreting causal necessity to mean causal dependence on something else; (iii) drawing the conclusion (with an equally perfect stringency) that the thesis is inconsistent with the implications of this interpretation. As long as there is neither a rational nor an empirical way of favouring one definition of 'causal necessity' against the other the deadlock will remain. Each side will dog-matically affirm that his definition is the 'true' or 'cor-rect' one which 'really' applies to the nature of things as they are in themselves. As long as we remain on this dogmatic plane the conflict can never be resolved.

[40] Observation on the antithesis.

Although the conclusion of the thesis is that the necessary being exists and forms an integral part of the world, still some Newtonians have tried to escape the latter part of the conclusion. As previously mentioned, Clarke, out of deference to orthodox Christian belief, tried to retain the belief that God somehow exists outside the world. Kant takes note of this point in the observation on the thesis where he explains that 'certain thinkers have allowed themselves the liberty of making such a *saltus* . . .'. These thinkers, not being able to find in the series of empirically determining causes a 'first beginning' or a 'highest member', have passed suddenly to an 'intelligible' first cause. The first cause is 'not bound down to any sensible conditions' and is 'freed from temporal conditions which would require that its causality should itself have a beginning'.[41] The conclusion of this 'entirely illegitimate procedure' is that the necessary being exists outside the world (outside space and time). Examples of this procedure can be found in certain of Clarke's declarations where he is doing his best, as a faithful Newtonian, to maintain an orthodox Christian view concerning the whereabouts of God. Kant summarizes this whole position in the *Dissertation* as follows:

Thus a whole of substances is a whole of contingent things, and the world, by its essence, consists of merely contingent things. Further, no necessary substance is connected with the world, save as cause is connected with effect, and not, therefore, as a part is connected with its complementary parts to form a whole. For the connection of joint-parts is that of mutual dependence, which does not apply to a necessary being. Thus the cause of the world is an extramundane

---

[41] Observation on the thesis.

being; consequently it is not the soul[42] of the world; its presence in the world is not local but virtual. (68)

Now P_4 in the proof of the antithesis is a refutation of the claim of those Newtonians who want to hold (as good Christians) that the necessary being exists outside the world. P_4 is not immediately directed to any explicit claim made by the thesis itself, for the thesis insists that the necessary being is in the world.

The argument of P_4 may be summarized and explained as follows:

1. Assume 'that an absolutely necessary cause of the world exists outside the world'.

2. This necessary being causes the series of alterations (forming the world) to begin.

3. To cause this series of alterations to begin the causality of the necessary being must itself have a beginning.

4. Since, according to the Newtonian definition, a beginning is always in time it follows that the causality of the necessary being occurs at one of the moments of infinite receptacle-time.

5. It follows that the necessary being (with its causality) is in time along with the world.

6. Hence: outside the world (space and time), there does not exist an absolutely necessary being which is in causal connection with the world.

In other words, the antithesis and its proof work to deny all possible interpretations of the Newtonian position about a necessary being whether this interpretation

---

[42] Leibniz accused Clarke and Newton of making God into something like the soul of the world.

makes him a part of the world or distinct from it. Thus the twofold proof of the antithesis concludes with the following statement:

Therefore neither in the world, nor outside the world (though in causal connection with it), does there exist any absolutely necessary being.

This argument of the antithesis tends in the direction of showing the Newtonians that they cannot hold that the necessary cause of the world (God) is outside the world and also that causality is meaningful only in terms of temporal precedence (relatively to the moments of absolute time). For this would reduce God's causality in the world to an unintelligible miraculous affair. It is precisely along these lines that Leibniz tries to face Clarke with a certain dilemma. The dilemma consists in putting Clarke before two unpleasant choices. On Clarke's assumptions: either God is a part of the world (i.e. He is corporeal) or His causality in the world is of the order of the miraculous and fantastic. Leibniz wrote to Clarke:

If God is obliged to mend the course of nature from time to time, it must be done either supernaturally or naturally. If it be done supernaturally, we must have recourse to miracles, in order to explain natural things: which is reducing an hypothesis *adabsurdum*: for, every thing may easily be accounted for by miracles. But if it be done naturally, then God will not be *intelligentia supramundana*: he will be comprehended under the nature of things. . . .[43]

The Newtonians also cannot have their cake and eat it too.

[43] Second letter, para. 12.

But if God is part of the world then He has parts. Leibniz raised this objection. He wrote 'but space has parts: therefore there would be parts in the essence of God'.[44] Kant reproduced this same argument against the Newtonians in the *Dissertation*. He wrote:

They imagine the presence of God to be local, and involve God in the world as if he were comprised in infinite space all at once. . . . But to be in a number of places at the same time is absolutely impossible, because different places are outside one another, and consequently what is in more than one place is outside itself and externally present to itself, which is self-contradictory. (77)

The conflict between the Newtonian and Leibnizian views on the question of a necessary being is clearly referred to by Kant in the *Dissertation* of 1770. Speaking from the point of view of 'the form and principles of the sensible world' (Newtonian space and time) he finds a demand for 'wholeness' in the sense of an *absolute* totality of conjoint parts' and not of a comparative totality of these parts (41). On the other hand, we have the opposing thought which Kant summarizes as follows:

For it can hardly be conceived how the never-to-be-completed series of states of the universe, succeeding one another to eternity, can be brought together into a whole comprehending absolutely all changes. On account of its very infinitude it is necessary that it be without limit, and therefore that no series of successive events be given save as part of a further series. Consequently, for the same reason, an all-round completeness, an absolute totality, seems to be ruled out altogether. (42)

---

[44] Fifth letter, para. 42.

Kant calls the opposition between these two ideas a 'thorny problem' (43). It is precisely this thorny problem which appears later on as the fourth antinomy of pure reason.

CONCLUDING REMARKS

I HAVE tried to represent the Kantian doctrine of the antinomy of pure reason as a basically successful attempt on his part to pit dogmatic metaphysical claims about the nature of reality against each other. His living example was the famous controversy between Clarke and Leibniz. In this controversy major traditional philosophico-metaphysical questions were subjected to vigorous discussion and reconsideration under the impact of modern scientific ideas (empiricism, materialism, atomism, mechanism, etc.) and the radical revisions they kept forcing upon the traditional and inherited picture of the world.

As such, my attempt is meant to be a reply to the claims of Norman Kemp Smith (and those who hold similar views) that: 'Kant fails to justify the assertion that on the dogmatic level there exist antinomies in which both the contradictory alternatives allow of cogent demonstration.'[1] Accordingly we should avoid imposing upon Kant standards of 'cogency' that are alien to his treatment of the subject. The sort of standards that are relevant in this context have to do with the 'cogency' of certain doctrines and assertions about the 'real' nature of the world 'which can neither hope for confirmation in experience nor fear refutation by it'.[2] Consequently, as long as the claims and arguments of each side of the antinomy are 'free from contradiction' then the counter assertion will have 'on its side, grounds

[1] *Commentary*, p. 519.
[2] B449.

that are just as valid and necessary' as the assertion itself.[3] The grounds produced by Kant in the proofs of
each thesis and antithesis are, more often than not,
versions of fairly well-known arguments in the history
of philosophical thought. These have been adduced as
'cogent' considerations (and at times as conclusive
'proofs') in favour of the finitude or infinitude of the
world as well as of other related doctrines such as the
antinomies deal with. This is why Kant insists that
antinomial conflicts cannot possibly arise either in the
pure formal sciences (logic and mathematics) or in the
empirical sciences (B452). They arise in the so-called
science of rational cosmology which claims to provide
synthetic *a priori* knowledge about the true nature of
things as they are in themselves. Similar considerations
apply to the so-called sciences of rational psychology and
natural theology.

The dogmatism of the opposed assertions in the antinomy arises ultimately from the claim of each party to
a superior 'insight' into the nature of things which
supposedly confirms its point of view and conclusively
refutes the opponent's. Such an assertion 'is therefore
itself dogmatic, claiming acquaintance with the constitution of the object fuller than the counter-assertion'
(A388). But for Kant 'the abstract synthesis' of such
'transcendental assertions which lay claim to insight
into what is beyond the field of all possible experiences
. . ., can never be given in any *a priori* intuition, and
they (the antinomial assertions) are so constituted that
what is erroneous in them can never be detected by
means of any experience' (B453).

I should note here that Kant's lengthy treatment of

[3] B449.

141

the doctrine of the antinomy is by no means limited to the strict epistemological and conceptual level that I have tended to emphasize. Mingled with all this are very interesting analyses of the psychological motives which lead people to stand for the thesis or the antithesis; plus detailed attention to the moral and practical implications of each thesis and antithesis. Kant investigates the moral interests that are at stake in accepting one or the other of the sides of the antinomial conflict. Added to all this is Kant's strong tendency to express himself in the language of faculty psychology. All these factors, though they add to the richness of Kant's treatment of the subject, tend to cloud the 'critical' and more interesting strains of his thought to the modern reader. For example, we need not be convinced, towards the end of the twentieth century, that man's inability to make any synthetic metaphysical assertions about the nature of things is a 'positive advantage to the demands of morality'[4] as Kant regarded them.

Another interesting point deserving mention is Kant's claim that there are four, and only four, possible antinomies that could arise before reason in its quest for knowledge (A462–B490). Commentators on Kant seem to be agreed that this claim is dictated by his faithfulness to the architectonic rather than by serious reflection over the problem. Practically every major and famous metaphysical dispute can be, and has often been, cast in the form of an antinomial conflict which can be resolved neither empirically, nor purely rationally. For instance the impasse reached between 'realism' and 'idealism' towards the end of the nineteenth century

---

[4] *The Bounds of Sense*, p. 215.

was treated by such philosophers as Whitehead, Bergson, Husserl, Russell, Dewey, and Sartre, as a deadlock before which no further appeal to experience or dialectical arguments can be of any use. This 'recent' antinomial conflict again served to awaken philosophy from its inherited dogmatic slumber and pointed to the irrelevance of the impasse for the continuation of the enterprises of philosophical thought. Accordingly Bergson went searching for the 'immediate data of consciousness', Whitehead invited thinkers to return to the revelations of our 'immediate aesthetic and moral intuitions', while Russell attempted to isolate the 'sensory core' present in all our sense perceptions.

Of course Kant himself was not as unaware of the possibility of formulating additional antinomies as his formal declarations imply. He mentions in this connection Zeno of Elea who:

. . . would set out to prove a proposition through convincing arguments and then immediately overthrow them by other arguments equally strong. Zeno maintained, for example, that God (probably conceived by him as simply the world) is neither finite nor infinite, neither in motion nor at rest, neither similar nor dissimilar to any other thing. (B530–A502)

Kant goes on to defend Zeno against his detractors and critics who accused him of being a mere sophist. Again, there is the metaphysical opposition, mentioned by Kant, between the materialism of Epicurus and the idealism of Plato. Kant presents this as an example of a generalized dogmatic metaphysical conflict about the nature of things as they are in themselves (A471–B499).

No account of Kant's doctrine of the antinomy can be considered adequate without some discussion of his proposed solutions to the conflicts. I do not need to deal with this part of the Kantian teachings in detail. P. F. Strawson has carefully examined the issues involved (in *The Bounds of Sense*) and made many profitable suggestions and interpretations. Furthermore, I shall not concern myself, in the following discussion, with considerations pertaining to the 'interests of morality' and faculty psychology which permeate Kant's statement of his views on the solution of the antinomial conflicts.

The general outline of Kant's solution to the problem of the antinomy may be profitably exposed in terms of a number of layers. First, Kant raises the question of whether the antinomy represents a 'genuine' conflict at all, i.e. a conflict between genuine synthetic claims about the world as the dogmatists, on each side, keep insisting. Kant suggests that the whole elaborate conflict might very well be due to no more than a simple, but common, misunderstanding between the two combating parties. The critical philosophical attitude would, in this case, consist in trying to locate the misunderstanding in preparation for removing it and then guarding against its recurrence in the future. Thus the conflict would, in a sense, disappear as a supposed serious and significant problem in the texture of our knowledge about the world.

This approach to the antinomial conflict (called by Kant the *sceptical method* as opposed to *scepticism*) watches the 'conflict of assertions, not for the purpose of deciding in favour of one or other side',[5] but in order 'to discover the point of misunderstanding in the case of disputes which are sincerely and competently conducted

[5] B451.

by both sides . . .'.[6] This approach leads to the following
beneficial results for human knowledge: (*a*) it shows the
futility of such dogmatic conflicts. This is why Kant
says that the vigorous fighters 'must be left to decide
the issue for themselves. After they have rather ex-
hausted than injured one another, they will perhaps
themselves perceive the futility of their quarrel, and part
good friends' (B451). Thus detecting the misunderstanding and clearing it up shows that both assertion and
counter-assertion in the antinomy are unsupported
rather than that they are wrong (A388). Furthermore,
this critical approach does not present any claims of its
own about the nature of objects after showing the base-
lessness of the antinomial claims. It simply 'confines
itself to pointing out that in the making of the assertion
something has been presupposed that is void and merely
fictitious . . .'. And it does not proceed to claim 'to
establish anything that bears directly upon the constitution of the object'[7] (A389).

(*b*) The clarification of the misunderstanding in question yields a net gain to human knowledge which takes
the form of realizing that 'the very fact of their (the
two disputants) being able so admirably to refute one
another is evidence that they are really quarrelling about
nothing . . .' (A501). Kant, again, characterizes the question over which the antinomial conflict occurs as 'null
and void'. He wrote: 'A question as to the constitution
of that something which cannot be thought through any

---

[6] B452.

[7] Kant says that the critical approach does not even claim that 'all
judgement in regard to the object (in itself) is completely null and
void'. For this would entail a claim to a special insight into the
nature of the object. This latter claim is shared alike by 'dogmatic
and sceptical objections' to the problem of the antinomy (A388–9).

determinate predicate—inasmuch as it is completely outside the sphere of those objects which can be given to us—is entirely null and void.' (A479–B507 n.) This shows the systematic irrelevance (and not necessarily the psychological, moral, cathartic, and regulative irrelevance) of the antinomial conflict to scientific knowledge and explanation of the world. Kant formulates this point as follows:

The absolute whole of quantity (the universe), the whole of division, of derivation, of the condition of existence in general, with all questions as to whether it is brought about through finite synthesis or through a synthesis requiring infinite extension, have nothing to do with any possible experience. We should not, for instance, in any wise be able to explain the appearances of a body better, or even differently, in assuming that it consisted either of simple or of inexhaustibly composite parts; for neither a simple appearance nor an infinite composition can ever come before us. (A483–B511)

(c) A certain beneficial effect of 'philosophical catharsis' results from all this which is described by Kant in the following terms:

By its means (the critical approach to the antinomial conflict) we can deliver ourselves, at but a small cost, from a great body of sterile dogmatism, and set in its place a sober critique, which as a true cathartic will effectively guard us against such groundless beliefs and the supposed polymathy to which they lead. (A486–B514)

The second layer in the critical discussion of the solution to the antinomial conflict is Kant's attempt to describe more specifically the nature and source of the misunderstanding underlying the antinomy. Working along these lines Kant characterizes the conflicting assertions (thesis and antithesis) as '*pseudo-rational* conclusions' and '*pseudo-rational doctrines*'.[8] They rest upon a certain illusion which makes the assertions look like rational and legitimate claims about the nature of things when they are not so at all. For Kant there are three types of such 'pseudo-rational inferences': 'the transcendental *paralogism*', 'the antinomy of pure reason', and 'the *ideal* of pure reason' (A340–B398). All their inferences involve what Kant calls 'the transcendental illusion'. Kant defines this illusion in the following words:

All *illusion* may be said to consist in treating the subjective condition of thinking as being knowledge of the object. (A396)

For an illusion to lead to the antinomies of pure reason (as well as to the paralogisms and natural theology) the assertion and counter-assertion:

. . . must involve no mere artificial illusion such as at once vanishes upon detection, but a natural and unavoidable illusion, which even after it has ceased to beguile still continues to delude though not to deceive us, and which though thus capable of being rendered harmless can never be eradicated. (A422)

Each antinomy falls into this illusion by regarding the opposed claims it embodies about the totality of things not as conceptual schemes of thought (which may be more or less useful: regulatively, psychologically, and morally) but as synthetic *a priori* propositions about the nature of the independently real.[9] In this sense, the antinomy treats certain ways of our thinking ('the

[8] B397, B449. Kant's italics.
[9] In this context Kant uses a variety of equivalent descriptions in referring to the independently real: 'things as they are in themselves', 'the transcendental object', and simply 'the object'.

subjective condition of thinking') 'as being knowledge of the object'. The antinomy fulfils Kant's definition of the transcendental illusion.

Kant also explains the nature of the transcendental illusions in somewhat different terms. The antinomy rests:

. . . on a mere delusion by which they (the conflicting dogmatists) hypostatise what exists merely in thought, and take it as a real object existing, in the same character, outside the thinking subject. (A384)

As a result of this tendency to hypostatize the conceptual there 'originates an imaginary science, imaginary both in the case of him who affirms and of him who denies, since all parties either suppose some knowledge of objects of which no human being has any concept or treat their own representations as objects . . .' (A395). For example, the extreme 'scientific realism' (called transcendental realism by Kant) of the Newtonians as embodied in the thesis of the first antinomy hypostatizes 'the transcendental idea of the absolute totality of the series of conditions in all past time',[10] as well as the idea of 'the absolute totality of appearances in space'.[11] The resulting doctrine is:

a *transcendental realism* which regards time and space as something given in themselves, independently of our sensibility. The transcendental realist thus interprets outer appearances (their reality being taken as granted) as things-in-themselves, which exist independently of us and of our sensibility, and which are therefore outside us—the phrase 'outside us' being interpreted in conformity with pure concepts of understanding. (A369)

[10] A412–B439.
[11] A413–B440.

In each of the other antinomies the dogmatists commit the error of hypostatizing certain 'transcendental ideas' which in themselves may be quite useful provided we do not claim for them the status of true assertions about the nature and constitution of the world as it is in itself.

This tendency to hypostatize our ideas is quite natural to man. Kant expresses this thought by saying that the transcendental illusion is 'natural' and 'unavoidable'. We should take care, I think, not to interpret this characteristic of the illusion psychologistically à la Norman Kemp Smith who speaks, in this connection, of 'an inborn need of metaphysical construction'.[12] The illusion is natural and unavoidable simply in the sense that no matter how meticulous and careful we are in our thinking about experience we cannot fully guard ourselves against hypostatizing our ideas. There is no way of completely eliminating the possibility of self-deception. The illusion can be detected and guarded against to a greater or lesser degree, but it cannot be eradicated. Kant, therefore, tells us that our only resort is the critical temper of the mind. He wrote: 'Nothing but the sobriety of a critique, at once strict and just, can free us from this dogmatic delusion, which through the lure of an imagined felicity keeps so many in bondage to theories and systems' (A395).

The third and final layer in Kant's treatment of the solution of the antinomies consists in turning the dogmatic assertions from presumed synthetic propositions about the world into 'regulative principles' or rules of procedure which are useful in the scientific investigation of phenomena.

The 'transcendental ideas' involved in the antinomies

[12] *Commentary*, p. 426.

are accordingly regarded as setting *a task*[13] before the investigator rather than as giving him final metaphysical information about the world. It becomes clear then that the mistake which Kant is labouring to expose consists in 'the ascribing of objective reality to an idea that serves merely as a rule' (A509–B537). For example, according to this line of thought, the question about 'the transcendental idea of the absolute totality of the series of conditions' (dealt with in the first antinomy) 'is no longer how great this series of conditions may be in itself, whether it be finite, or infinite, for it is nothing in itself; but how we are to carry out the empirical regress, and how far we should continue it' (A514–B542). The idea of the cosmic whole can only serve as a rule of procedure which:

says no more than that, however far we may have attained in the series of empirical conditions, we should never assume an absolute limit, but should subordinate every appearance, as conditioned, to another as its condition, and that we must advance to this condition. (A519–20, B547–8)

Beyond that, reflection over the idea of a cosmic whole cannot yield any information on whether the world in itself is finite or infinite.

Accordingly Kant praises Epicurus by describing him as possibly the wisest of the philosophers of antiquity. Epicurus seems to have treated his materialistic doctrines not as a strictly true account of the nature of things but as a useful system of concepts which helped him in understanding phenomena. Kant wrote a passage in which he praised Epicurus and laid the foundation of the 'philosophy of the as if':

It is, however, open to question whether Epicurus ever

[13] A498.

propounded these principles as objective assertions. If perhaps they were for him nothing more than maxims for the speculative employment of reason, then he showed in this regard a more genuine philosophical spirit than any other of the philosophers of antiquity. That, in explaining the appearances, we must proceed as if the field of our enquiry were not circumscribed by any limit or beginning of the world; that we must assume the material composing the world to be such as it must be if we are to learn about it from experience; that we must postulate no other mode of the production of events than one which will enable them to be, [regarded as] determined through unalterable laws of nature; and finally that no use must be made of any cause distinct from the world—all these principles still [retain their value]. They are very sound principles (though seldom observed) for extending the scope of speculative philosophy. . . .' (A471–B499 n.)

According to this line of thought the idea of the finite or infinite divisibility of matter involved in the second antinomy becomes the simple rule of procedure that 'in the decomposition of the extended, the empirical regress, in conformity with the nature of this appearance, be never regarded as absolutely completed' (A527–B555). Dogmatic claims about the finite or infinite divisibility of matter will lead us nowhere.

Again, in the case of the third antinomy, the idea of ascribing or denying to 'agents' 'the power of originating a series of events' is to be treated as a 'regulative principle of reason'. We can 'illustrate this regulative principle of reason by an example of its empirical employment', but we cannot prove it 'for it is useless to endeavour to prove transcendental propositions by examples . . .' (A554–B582). Kant's example of this 'empirical employment' is a simple voluntary action, the telling of a malicious lie which causes confusion in society. The regulative principle requires us to think and behave in this case 'just as if the agent in and by himself began in this action an entirely new series of consequences' (A555–B583). On this basis only can we hold him responsible for the lie. Contrary to the dogmatic claims of the thesis and antithesis we can have no assured synthetic knowledge of whether such 'agents' are in themselves free or not. Kant holds on this point:

The reader should be careful to observe that in what has been said our intention has not been to establish the *reality* of freedom as one of the faculties which contain the cause of the appearance of our sensible world. . . . It has not even been our intention to prove the *possibility* of freedom. For in this also we should not have succeeded, since we cannot from mere concepts *a priori* know the possibility of any real ground and its causality. . . . What we have alone been able to show, and what we have alone been concerned to show, is that this antinomy rests on a sheer illusion, and that causality through freedom is at least *not incompatible with* nature.[14] (A558–B586)

Norman Kemp Smith clarifies Kant's idea in this passage by saying that 'the only point established is that freedom is, so to speak, a *possible possibility*, in that it is *not contradicted* either by experience or by anything that can be proved to be a presupposition of experience'.[15] Strictly speaking, thoroughgoing determinism is also not contradicted by experience or any presupposition of experience.

The idea of a necessary being, as dealt with in the last antinomy, is also turned by Kant into a regulative principle which is useful in the investigation of empirical

[14] Kant's italics.
[15] *Commentary*, p. 517. Smith's italics.

phenomena. It cannot form, either on the side of the thesis or the antithesis, a synthetic statement about the nature of things. Kant says:

The *regulative principle of reason*, so far as it bears upon our present problem, is therefore this, that everything in the sensible world has an empirically conditioned existence, and that in no one of its qualities can it be unconditionally necessary; that for every member in the series of conditions we must expect, and as far as possible seek, an empirical condition in some possible experience; and that nothing justifies us in deriving an existence from a condition outside the empirical series or even in regarding it in its place within the series as absolutely independent and self-sufficient. (A561–B589)

Naturally, it remains open for any one to believe (for moral or psychological reasons) that freedom (absolute spontaneity) and a necessary being exist in the realm of intelligible objects. This belief is 'an optional assumption'[16] in so far as the 'real' existence of freedom and of a necessary being can be neither confirmed nor disconfirmed by any possible experience.

[16] A562–B590.

INDEX

INDEX

Leibnizian, 5–6, 8, 20–2, 26–8, 45, 52, 58, 62–3, 66, 74–5, 81, 84, 89, 119, 133, 138.
Locke, J., 12–13, 19.

Martin, G., 1, 8, 24, 113.
materialism, 4, 116–20, 130, 140, 143, 150.
matter, *see* substance, material.
monads (monadists), 7, 66, 72, 80–1, 83.
motion, 25, 50, 52, 56, 81–3, 95–6, 99, 118, 143; absolute, 24, 26, 81.

necessary being, 112–16, 119, 120, 122–8, 131–3, 135–8, 152–3.
necessity, 126–8, 133, 134.
Newtonian, 3, 7–9, 14–17, 20–2, 24–5, 29, 33, 35, 43, 52–6, 58, 61–3, 66–8, 75, 78, 81–3, 94, 116–19, 124–5, 129, 131, 133, 135–6, 138.

ontological argument, 114.

Paulsen, F., 4, 115, 119.
Plato, 4, 143.
Prichard, H. A., 8.

rationalism, 3, 29.
realism, transcendental, 148.
regulative, 91, 149, 152, 153.
Russell, B., 143.

Sartre, J. P., 143.
Schrader, George, 8, 21.
self, *see* substance, mental.
simple, the, 10, 12, 48–51, 54–8, 60–1, 64–5, 68–72, 74, 79–80, 146.
simple location, 54.
Smith, Norman Kemp, 1, 8, 17–

20, 24, 46, 51–2, 55, 73–4, 83–4, 112–13, 140, 149, 152.
soul, *see* substance, mental.
space, 2–3, 6, 8–9, 11, 13, 17–19, 21–8, 32–4, 36, 38, 41, 50–3, 61–3, 66, 75–82, 116–17, 121, 125, 127, 130, 136, 138, 148; absolute, 4, 14, 23, 25, 34, 41, 52, 55, 63, 78, 81–2, 84, 117–19, 129.
Spinoza, 116, 118, 130.
spontaneity, *see* freedom.
Strawson, P. F., 14–16, 43, 48, 61, 144.
substance, 46–50, 53–4, 56–8, 61, 64–5, 68–72, 80–1, 84, 96, 109–10, 125, 132, 135; material, 6, 46–7, 50–3, 55–7, 59–63, 65–7, 73, 75, 78–9, 82–5, 96, 116, 124, 151; mental, 46–7, 56, 59–60, 67, 73–4, 87, 96, 116, 136.
sufficient reason, law of, 25, 27–8, 30–5, 45, 56–7, 63–4, 71–2, 87, 107–8, 134.
synthesis, 9–13, 16, 19, 20, 43, 141, 146.

time, 2–3, 6, 8–9, 11, 15, 17, 19, 21, 22–3, 27–8, 34, 38, 41, 42–5, 50–2, 82, 89, 97, 114, 116, 121, 125, 127–8, 130, 136, 148; absolute, 41–5, 67, 117–19, 126, 129, 137.
totality, 9–13, 15–19, 22, 25, 28, 40, 50, 93, 122–5, 127, 131–5, 138, 146–8, 150; synthetic, 11, 13, 18–19, 51, 93, 122; analytic, 11, 13, 16, 18–19, 50–1.

universe, material, *see* world.

vacua, 24, 26.
Vleeschauwer, de, 1.